Techniques in Engineering Design

G. PITTS B.Sc. (Eng.), Ph.D., C.Eng., M.I.Mech.E., M.I.Prod.E.

Lecturer in Mechanical Engineering Design
University of Southampton

A HALSTED PRESS BOOK

JOHN WILEY & SONS

New York - Toronto

Published in the U.S.A. and Canada by Halsted Press
a Division of John Wiley & Sons, Inc., New York

Library of Congress Cataloging in Publication Data

Pitts, G.

Techniques in engineering design.

"A Halsted Press book."

Bibliograhy: p.

1. Engineering design. I. Title.

TA174.P57 620'.004'2 73-11242

ISBN 0-470-69077-1

Printed in Hungary

Preface

The scope of engineering design is unbounded; the intention of this work is not to attempt an all embracing thesis on the subject, but to outline basic design philosophy and to relate some currently applied techniques to this design process.

The aim is threefold. First, to provide a useful text for undergraduate degree courses—the author's own experience indicates that no other book has dealt with the subjects of optimisation, curve and surface definition and computers in design in a simple and concise way suitable for undergraduate design teaching. Secondly, many practising designers have little or no opportunity to study modern design methods and techniques, and most are unaware of the true role of the computer as a design tool. Finally, students on other technical courses should find this book useful, either in part or as a whole, for supplementary reading.

The first four chapters will provide easy reading for all levels of theoretical attainment. Chapter five requires a reasonable standard of mathematical understanding but at the same time provides an ideal opportunity to consolidate earlier studies of vector notation in a graphic way. Chapter six demands an elementary knowledge of mathematics linked with a fairly logical mind. Many of the examples are of an open ended nature, which typifies design. In the early chapters the reader is encouraged to be observant and to check his finding against everyday design solutions. Towards the end of the book many of the answers can be tested using an alternative approach; these are necessary in real design situations.

Preface

This book could not have been written without the help and encouragement of many people, in particular: Mr A. J. Munday, for his constructive observations on the manuscript; also Professor S. P. Hutton, Mr C. H. Helmer, Mr K. D. Sadd and other members of the University of Southampton Design Group, for helping to create an environment of design thinking and logic; a patient publisher; Mrs Arnold-Jenkins for her many hours of typing, retyping and correcting; and, last but most important of all, my wife and young son, though he knows it not, for so arranging their lives to make its writing possible.

G.P.

Contents

1 Introduction

1.0 THE INSIGNIFICANT DESIGNER

In a majority of industries a designer's influence is comparable with that exercised by management. Every viable industry must work to an economic yardstick and since design office expenditure is small compared with that of manufacturing units, it is often mistakenly supposed that this relates to the significance of the designer. In fact a similar argument could be levelled at management, which is clearly untrue. How then can a designer exert this influence?

His most common form of communication is the drawing; every single line on a drawing is a manufacturing instruction, and every instruction requires both time and money to fulfil. Until the last decade any design capable of being manufactured, having left the design office, has normally gone undisputed through the production unit regardless of time or cost. It is only recently that value analysis has proved the designer's human fallibility. This power has been further enhanced by high production rates, e.g. a badly designed small automotive component may cost the company millions of pounds, especially bearing in mind the increase in spares variety which could result from redesigning. It is often said that a well designed product sells itself; whilst this is not literally the case, a well designed product can increase the market potential well beyond the achievements of the most ambitious sales promotion. Only a few of the ways in which design can influence a company's prosperity have been considered; it now remains to be shown what is required of a good designer.

More specifically, what is meant by a designer? One often hears the expression, 'Oh he's not a designer he's just a detail draughtsman'. Any engineer who has to create a solution to meet a specific requirement is designing. Detail design is just as significant as the overall design; unreliability nearly always stems from a failure in detail, and manufacturing costs are influenced as much, if not more, by the attention to detail. By virtue of his far-reaching responsibilities the essential qualities of a good designer are comparable with those of the manager and, as with management, they are demanded at all levels of design.

The most important attribute is that of logical decision-making. Every step in a design requires decision—decision not based on prejudices and feelings, but on available facts. The difficulty with any important decision is that as far as possible it must be correctly arrived at and then adhered to. A limited availability of facts can only lead to doubtful decisions. Usually the missing facts are not those which are central to the problem but those on the problem periphery. For example the designer may satisfy the requirement but he may be unable to judge the customer reaction to the finished product. Therefore, decisions must be broadly based. There is, of course, the old adage that decisions, like laws, are made to be broken. Far too often the designer faced with a problem finds an 'instant' solution and blindly follows it or, in contrast, realises that the perfect solution does not exist and continues to waver indecisively. A good designer must always be able to draw the fine line between perfection and acceptable compromise.

All sound decision making depends on recognising the problem in its basic form, eliminating superfluous trimmings which tend to confuse the issue, and assessing it objectively. Frequently there is unnecessary attention to the irrelevancies and little or no attention to the basic requirement.

A pure research worker can afford to disregard the ultimate outcome of his work, and is free to follow the direction he fancies most. This is a luxury eluding the designer who must at all times be working towards a marketable commodity, available on the target date and at a price the market will stand. At the same time he must be an innovator and pace setter providing a bridge between fundamental research and engineering production. As no man can be a universal expert it is necessary to use the expertise of others and to know where advice can be found when required. An ability to communicate is probably one of his most important qualities. Not only must he communicate with those from whom he seeks ideas but he must be able to transfer the products of his skills to the manufacturer. The engineering drawing in its various forms has, for centuries, been the principal mode, but the current trend is to eliminate much of the routine drawing work from the design office. Many companies are encouraging the use of freehand sketches to describe simple components, and the aircraft industry has transferred all detailing, which will ultimately require templates, to the loft where templates are drawn directly from assembly drawings. It is now realised that verbal or graphical communication are at times the most effective means of transferring information. The advent of numerical control has introduced the magnetic tape as a further transfer channel.

Finally, many designers look upon themselves as mechanical, electrical or civil engineers and have decided their points of demarcation. This is a contradiction of the basic requirement that a designer's outlook should be broad rather than deep. Far too often he is aware of a possible solution a little outside his own field, and rather than explore new territory he will content himself with an inferior solution on a familiar stomping ground.

1.1 WHY DESIGN TECHNIQUES?

Many good designs have, in the past, been arrived at in one of two ways. In the first there has been the so called 'natural' designer, who appears to have a 'flair' for producing a sound design at almost every attempt. In fact what is really meant is that such a person is a logical and tidy thinker ready to explore all possibilities, arriving at the optimum after a careful evaluation of the alternatives. Unfortunately, the second way involves a chance encounter with the most appropriate solution to the problem and whilst the designer may bring forth one good design, his other products may be acceptable but poor.

Modern technology has placed heavy responsibilities on the designer. Half a century ago an experienced mechanical engineer could draw upon his experience to produce a most advanced technical design of the day, and in so doing would be required to study possibly half a dozen alternatives all within the mechanical field. Currently, he may need to consider fifty or more alternatives before reaching a decision, many of which would fall outside his own experience. Consider for example the design of a precision measuring device. At the turn of the century the options open to a designer requiring to magnify a measurement were limited to mechanical arrangements such as a screw micrometer, a rack and pinion (dial-gauge principle) or a simple lever, etc. Currently his options would include: the mechanical methods, electronic amplification, optical levers, optical interference, coherent optics, fluid logic devices, pneumatics, radiation methods, etc. A designer working in the measurement field must acquaint himself with all these alternatives, realise their potentials and know something about their availability and cost. As the years tick by the options gradually increase. With such scope the designer can no longer manipulate these alternatives in his own mind but must employ modern evaluation techniques to reach an orderly decision.

The evaluation techniques fall into two general classes: general decision methodology, and mathematical comparisons between calculated values, described in this text as optimisation. The engineering drawing as a method of graphical communication has been established for centuries, but the present day designer must know something of the science and craft of communicating by numerical means.

2 The design process

2.0 DESIGN, CREATIVITY AND INVENTION

What is meant by design? The titles: designer, inventor and draughtsman, tend to suggest to most people an individual's status rather than his function. The greatest difficulty occurs in differentiating between design and invention, if indeed there is any difference! It is far better to examine the underlying need for creation.

Creation generally arises for one of two reasons: either an individual comes upon a chance idea which when exploited could benefit others, or alternatively the individual is faced with a problem and must attempt to solve it. In the former situation one may describe the person as an inventor, although strictly speaking an inventor is better defined in the language of the patent agent. What is found on further examination is that the creator becomes interested in a particular field of activity and allows this interest to stimulate ideas.

Pye[1] describes this form of creation as creation stimulated by play. It is interesting to note that these 'inventors' rarely limit their work to one particular field but hold patents on products spanning a whole range of industries.

In practice the engineering designer cannot allow his fancies to wander over the different fields of engineering in this way, his work is much more limited. He is presented with a problem to which an economic solution must be found within a reasonable period of time. The way in which he sets about finding this solution should involve a logical design process to satisfy a specific requirement.

2.1 SYSTEMATIC DESIGN

If the designer is to proceed from the problem to the solution in a logical manner he must first understand the mental process which is carried out in the creation of a good design, and then discipline himself to follow it at all times. The mental process must of course be related to the physical stages of design. In training a designer to work to a system it is essential that he commits some of this system to paper. It may be done as part of his early training on simple problems, to be continued when the problem is large and the possible solutions extremely diverse, the simpler jobs can then be handled by his 'natural flair for design' or more correctly subconscious following through of a logical process.

All design must start with a need or requirement, from this the designer decides what the objectives are. Having established the objectives, the best possible way of meeting these must be sought, and finally the design must be engineered to the stage of being a manufacturable product. This philosophy is essential to all decision-making processes.

The stages in the design process may be listed as follows:

(i) Requirement or designer's brief
(ii) Problem specification
(iii) Search for alternatives (speculation)
(iv) Evaluation of alternatives
(v) Synthesis

Clearly steps (i) and (v) are unavoidable in any design, but stages (ii), (iii) and (iv) are the mandatory ingredients to a sound result. A common fault is to proceed directly from the requirement to the synthesis, overlooking some of the more fruitful possibilities.

2.2 REQUIREMENT

The requirement outlines the need to be met. The first point to be considered is the reliability of the source from which the requirement originates, this may be internal or external to the designer's own organisation. Internal requirements generally originate in one of two ways, either as a subdivision of a larger design, e.g. the consequence of designing an engine is that mountings must be designed for it, or alternatively a potential market has been seen by the company and hence the new product. Designs originating as subdivisions of larger systems form the bulk of design office work. Unfortunately, there is often a failure to recognise these jobs as designs in their own right, and merely to consider them as an appendage to some larger system. External requirements are presented to the company by a customer and may reach the designer second or even third hand via the company's marketing department.

A major difference between internal and external requirements is the difficulty of assessing the expertise and reliability of the external source. The customer may be aware of the problem and be able to point to the need but may not have the specialist knowledge to define the need exactly.

When a doubtful requirement arises it must first be justified. Frequently a customer is unaware that a suitable product already exists, meaning that a new product must have technical, economic or political advantages to justify its design and manufacture; although the customer may be content to buy the product initially, on discovery of the cheaper and better item, policy may rapidly change. A further consideration is the customer or salesman realising a need for a product to satisfy a limited market at a specified maximum price. The designer must then assess the product cost in the light of the numbers required. Only if the cost is likely to fall below the specified maximum is the requirement justified.

This raises a further point which should always be paramount in a designer's mind when studying a requirement: Can the potential market be expanded? Customers are frequently unaware of the needs of others in similar fields. Often a requirement may be expanded slightly without an appreciable increase in complexity, resulting in a greater market potential at a reduction in price to the original customer. An example is a cave rescue hoist, the customer being a rescue organisation. A similar requirement may exist in mine, cliff and mountain rescue, in which case the market may be expanded tenfold, being the difference between a very expensive, individually made hoist and one produced using batch production techniques at a lower cost.

Summarising:

(i) Check the reliability of the requirement source.
(ii) Do not fail to recognise a subdivision of a larger design as a requirement in its own right.
(iii) Make a preliminary assessment of the requirement to see if it is economically realistic.
(iv) Can the requirement be simply modified to expand the market potential?

2.3 SPECIFICATION

A specification is a listing of all the parameters essential to the design.

The design specification must always be formulated by the designer, it can never come from the customer. A critical look at the problem in hand is a necessary first step in creating a correct attitude to the design. A customer may list essential values as a part of his requirement, but each value must be examined carefully and before transfer to the specification. Many of the statements made in a requirement are irrelevant to the design and often misleading. A trivial example might be a request for a wooden tool cupboard. The real need is for a tool cupboard; by stating wooden the customer has attempted to take a design decision. The specification should read: an enclosed form of tool storage with easy access. The designer must decide why the customer stated wooden. Is it because of appearance, then the specification might read: sturdy appearance, or is it because tools do not clatter on wood and that condensation problems are eliminated. The specification should then be a material with high vibrational damping properties and good thermal insulation. For a cupboard to be produced in quantity, a hard plastic might prove more satisfactory.

This leads to a second point: a specification must be so framed as not to influence the final solution, other than the solution satisfying the specification. The specification must be a factual statement of the conditions to be satisfied in the requirement.

There are two ways in which the items making up the specification may be presented. First, there are those characteristics which may be quantified, usually as a result of calculation. The quantifiable items are normally physical characteristics e.g., output power required, ambient temperature and pressure variations, the load which a structure must support, frequency of use, expected life, reliability, etc. In the second category are characteristics which in themselves cannot be quantified, but which may be described precisely and briefly. Many of the subjective aspects of the specification appear in this class, e.g. appearance related to the potential market, but there are also objective aspects, e.g. user skill, safety standards and service requirements.

It is rare to find that all items of the specification are of equal significance. An example of contrasting priorities would be a vacuum cleaner component and one required for an aircraft control system. It may not be possible to meet the reliability requirements of the vacuum cleaner within the acceptable price range, the price would naturally be the more significant feature of the specification. In contrast the aircraft control component would need to be made at increased cost to meet the reliability requirement. Matousek[2] suggests a graphic way of presenting the items of the specification, providing a vivid indication of the significance of each. In Figure 2.1(a) items from the vacuum cleaner specification are compared with those of the aircraft control component in Figure 2.1(b). A listing of items in this way may appear laborious, but it does provide the designer with a reminder of his priorities and some of the effort can be eliminated by providing a standard layout blank, requiring only the listing of items and the freehand shading of the priorities grid.

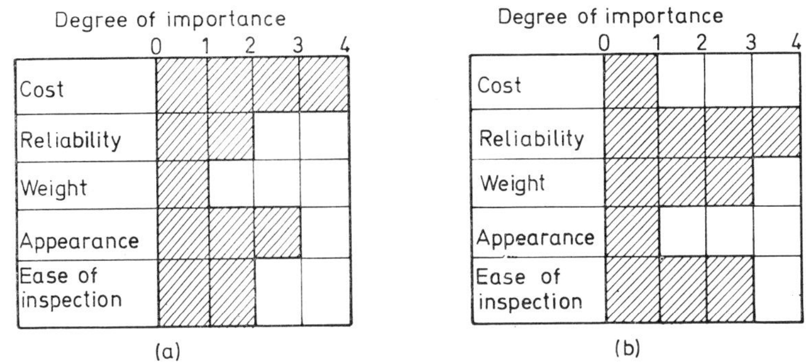

Figure 2.1 (a) The importance of some items from a vacuum cleaner component specification; (b) The importance of some items from an aircraft control component specification

Finally, incorrect specifications are one of the prime causes of over design and of redundant features appearing in the finished product.

Summarising:

(i) The specification is a listing by the designer of parameters essential to the design.
(ii) Any item or value transferred from the requirement to the specification must be critically examined before the transfer is permitted.
(iii) The specification must be so framed as to avoid unnecessarily influencing the possible range of solutions.
(iv) An order of priorities within a specification must be recognised.

2.4 THE SEARCH FOR ALTERNATIVES (SPECULATION)

Generating ideas comes from stimulating the imagination. What kind of stimulus is the designer likely to encounter and how can it be developed to bring forth a greater harvest of alternatives? In most situations the fact that the designer is presented with a new problem is sufficient in itself to suggest at least one method of solution. This results from an association between the problem and something the designer is able to recall from his own experience, and tends to follow a pattern of a standard solution to a standard problem. It is all very well to handle problems in this way provided that the original solution was the best available, and that circumstances have changed little in the meantime. Many designers are inclined to follow this system, and it is frequently said that experience is the only important quality required of a designer; a little thought will show this to be untrue. Although experience associated with the job in hand is one valid source of stimulus, it is very limited and it is necessary to consider other means by which the imagination may be extended.

There are basically two forms of stimulus. The first relies on human interaction and has come to be known as brainstorming. This is based on group discussion resulting in the reactions between members of the group generating new ideas. A second method is to develop a particular path of logic and allow this to expand the alternatives. The process normally begins at a somewhat abstract level before proceeding to the stage of engineering hardware.

Brainstorming originated in advertising as a way of discovering new methods of promotion. An essential ingredient to brainstorming is a relaxed, informal and somewhat jovial atmosphere (a point effectively made by Harrisberger[3]); this may be taken to the extent of easychairs and light refreshment provided at company expense. The participants are first familiarised with the problem and with any associated issues. Brainstorming proper then begins with the first idea for a solution being uttered by one of the group. This triggers another idea from a second member of the group and so on. An important rule is that all ideas must be recorded without any discussion taking place on their viability, and that participants are encouraged to throw up any 'ridiculous' possibility which enters their head. In this way the response is uninhibited and many of the 'ridiculous' ideas prove to be sensible, and if they are not, they have still provided a trigger for more realistic proposals. When all the possibilities have been exhausted the alternatives are carefully assessed and discussed in detail.

Brainstorming is a useful technique now extensively practised as part of value engineering, but it has two drawbacks. An obvious difficulty is the gathering together of a group at a specific time. The bulk of design results as a consequence of other design work. It is not uncommon for a designer to have to speculate several times in one day and to arrange brainstorming groups at such frequent intervals is not a very practical proposition. As a consequence, brainstorming is normally reserved for the major problems and for those where a solution eludes the designer. A second difficulty is keeping the specification accurately in mind and still maintaining an informal atmosphere. Situations can occur where the line of thought is allowed to drift away from the specification to such an extent that, whilst most of the solutions may satisfy the requirement, they include redundant features not appropriate to the specification.

The alternative method of establishing a logical path can be started by considering basic physical concepts. It has the advantage of involving only the individual designer, but at the same time relies heavily on the skill of the designer in its successful exploitation. The principle will be shown by an example. A transmission system is required to link the two shafts A and B shown in Figure 2.2. The axis of A is at 90° to the axis of B and their axes lie in different planes. The specification would give their relative positions, the power available at A, the torque or speed of A, the power available at B, hence efficiency, etc. A brainstorming approach may throw up several useful alter-

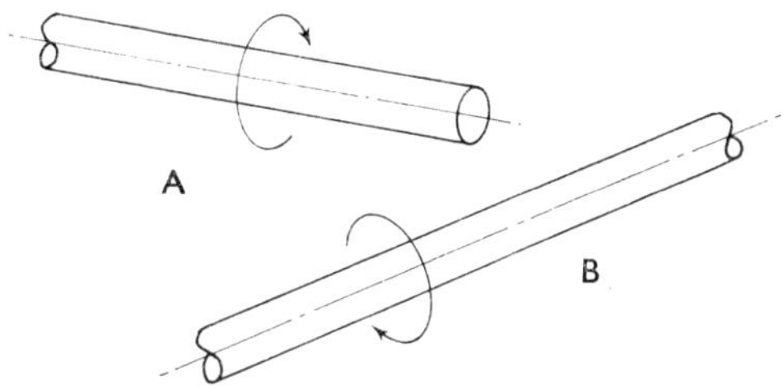

Figure 2.2 The problem of providing a drive between mutually perpendicular shafts with their axes in different planes

natives, possibly: extend A and use a twisted belt, chain and bevel gears, belt and bevel gears, hypoid gear, worm and wheel, etc.

The physical requirement is that power must be transmitted from A to B. The alternative ways of transmitting power are: mechanical, electrical, thermal, chemical.... In a typical situation given by Figure 2.2 it is likely that the mechanical method would provide more feasible alternatives than the others. The mechanical problem may be subdivided into two steps, the first is that of transmitting the power through an angle of 90° and the second is translating the power between the two parallel planes containing the axes of A and B. There are two types of solution: the complete solution, and the two-stage solution satisfying the need for translation and 90° rotation. Some examples are listed in *Table* (2.1).

Table (2.1) MECHANICAL POSSIBILITIES FOR THE DRIVE SHOWN IN FIGURE 2.2

Complete solutions	*Partial solutions*	
	Translation	*90° rotation*
Flexible drive	Spur gears Friction wheels	Bevel gears Friction cones
Worm and wheel	Chain drive	
Hypoid gear	Belt drive	
Twisted belt	Toothed belt	+ Some of the solutions listed in the left-hand column
Friction wheels		

For a complete solution any two of the partial solutions may be combined, provided that they are compatible. A range of mechanical alternatives has been generated, which would be extended by considering electrical methods if these were realistic, thermal and chemical ways may be expanded in a similar way, although this would at first sight appear fruitless. In many cases, it is worthwhile considering even the way-out possibilities. Many a designer has tried transmitting small powers through the side of a sealed vessel, employing elaborate mechanical seals for the purpose, when the transmission might well have employed a simple magnetic field at the interface!

Because of the difficulty encountered in expanding the range of alternatives, many of the formal design methods are formulated to provide the necessary stimulus.

Summarising:

(i) Some form of stimulus is necessary to generate ideas.
(ii) Confrontation with the problem is in itself a form of stimulus, but this is inadequate for most design purposes.
(iii) The two main forms of stimulus applicable to good design are: human interaction (brainstorming) and the pursuit of a logical path.

2.5 EVALUATION OF ALTERNATIVES

The exercise of generating alternatives becomes rather pointless if the merits of the resulting solutions are improperly assessed. At the same time, a detailed assessment of each alternative in turn, which unfortunately often occurs, would only result in a waste of the designer's time. It follows, therefore, that the alternatives must be assessed in stages, until the promising few are studied in detail.

The first stage of assessment can be made on the basis of common sense. Many of the solutions resulting from discussion may be of a ridiculous nature and have only served to provide more realistic ideas. Such solutions as skyhooks or 'tie it with string' are clearly in this category, although they may have led to more realistic propositions such as: support the structure by helicopter during assembly, and use a cable rather than something rigid since all the loads are tensile.

The next stage in the elimination is one of considering technical feasibility. At this level the designer is only interested in orders of magnitude to determine whether certain solutions are hopelessly out of the question; in the example just quoted it is unlikely that an electric generator–motor combination could have been accepted because of the large space required.

Many designers unwisely proceed to eliminate the alternatives on technical grounds alone. The correct approach is to follow each round of technical asessment with a corresponding level of economic assessment, becoming more detailed as the evaluation proceeds. If the generator–motor combination had not failed on the initial technical assessment, it would have very soon collapsed in the following economic assessment. In general more possibilities are eliminated on economic grounds than technical, and as a consequence systematic economic–technical considerations provide a more rapid convergence towards the final result. It is rare to find that the few remaining solutions are so closely competitive that a detailed analysis has to be made before arriving at a final decision. Other factors become more dominant, e.g. infringement of patents, avoidance of direct competition.

It is during the evaluation stage that approximate design charts, such as the Engineering Sciences data sheets[4] and approximate formulae are most useful in providing rapid comparisons; detailed calculations are deferred for the synthesis stage.

Needless to say, if all possibilities are eliminated then a commercially viable solution does not exist. On the other hand if economic considerations are excluded and a solution remains then a technical solution is possible which has proved to be commercially uneconomic.

Summarising:

(i) Careful assessment avoids throwing away the effort made in the earlier stages of the design process.
(ii) Evaluation should start at a superficial level, increasing in detail as the elimination proceeds.
(iii) Economic as well as technical considerations should be assessed at each stage of the elimination.

2.6 SYNTHESIS

Synthesis involves the engineering of the design to the state of preparing working drawings, component schedules, and other media of communication between the designer and manufacturer. It is here that a knowledge of manufacturing techniques, comparative costs and experience from previous designs becomes essential.

The nature of synthesis varies according to the complexity of design. At one extreme, the design of a simple component may demand a single drawing. In contrast, synthesis of a large system may generate new requirements for each subdivision; in such cases each requirement becomes a design study in its own right. Perhaps one of the greatest dangers in design synthesis is this failure to recognise new design requirements as and when they arise, leading to a requirement being followed immediately by an attempt at synthesis—a fault outlined in section 2.2.

Synthesis usually starts at a relatively conceptual level where the basic decisions on approximate cost and performance has been taken between alternatives. The first stage of synthesis may require detailed calculation of certain quantities, based on values presented in the specification, e.g. local load distributions in a framework, or speeds of rotation for intermediate shafts in a transmission system. Once these are established, the physical layout of the design can commence, including the relative positioning of the individual components within the assembly, their shape, component materials, methods of joining and linking, etc., culminating in detail drawings of the component parts.

The job of detail design is often given to the 'new boy' with the mistaken impression that all the design work has been completed and what is required is a 'pretty picture' for transfer to the workshop. The difference between the various levels of design is not one of skill, but one of changing emphasis. The qualities required at the initial stages are those of grasping basic physical concepts; a detailed knowledge of manufacturing processes although necessary may be given less emphasis. At the detail design stage the reverse is the case. More projects fail on inadequate attention to detail than from the lack of initial inventiveness!

Summarising:

(i) Be aware of new requirements arising during synthesis and handle them as designs in their own right.
(ii) Synthesis is the correct place for detailed analysis, not the evaluation stage.
(iii) Detailing is as important as the other design stages, and requires a specialised knowledge of manufacturing methods.

Although the stages in the design process have been laid out as discrete steps, an interaction must exist between them. Design is an iterative process, particularly on large projects, and it is not uncommon to find that major design modifications have to be made to facilitate the economic manufacture of individual components, nor is it uncommon to downgrade a minor item in the specification to obtain a commercially viable product.

2.7 THE ROLE OF DESIGN METHODS

In recent years many systems have been devised to assist in a systematic approach to design. These methods discipline the designer to follow the full design procedure and, when implemented within a design department, ensure a degree of uniformity between individuals. With some, the documenting of the design procedure enables an engineer handling the job at a later date to be aware of the designer's reasons for the particular solution he selected.

Three such methods will be outlined in the following sections and related to the design process. Their advantages and disadvantages will be examined. It is not suggested, however, that these particular methods are superior to the many similar routines which are to be found, but provide typical examples of how the design process is implemented in practice. Indeed, the reader may be able to devise a system more appropriate to the needs of his own work. It is difficult in the space available to consider each method in detail, for a greater understanding the quoted references should be consulted.

2.8 PABLA (PROBLEM ANALYSIS BY LOGICAL APPROACH)

PABLA was devised by Latham and Terry[5,6] for use at the Atomic Weapons Research Establishment, Aldermaston. Its use has been extended to other organisations. The system, based on methods study practice, is devised for the design draughtsman. It attempts to lead the user through the full design process by having him complete the standard PABLA forms, leaving a well documented line of design thought, at the same time creating the correct attitude to the job. The originators recognised the problem of the designer's inclination to take up a drawing pencil at the first opportunity; completing the charts provides an outlet for this enthusiasm whilst allowing the user to think in depth about the problem.

The method is best described by reference to the charts shown in Figure 2.3. C1 and C2 together cover the specification, whilst chart C3 handles the alternatives and their evaluation, C4 providing the necessary data for synthesis.

The box titles on the first chart are really posed as questions: occasion—when is the product used; duration—for how long; frequency—how often; sequence—what causes it to be used, etc.? Hence this chart provides the basic data essential to the specification. Two criticisms may be levelled at its layout: the first is the rather illogical numbering of the boxes from the user's point of view, bearing in mind that most users would want to complete the chart column by column, and the second is the inclusion of the third column at this stage in the operation. The existing resources column has the advantage of providing a dumping place for the designer's knowledge of solutions to similar problems but if he uses this column as an invita-

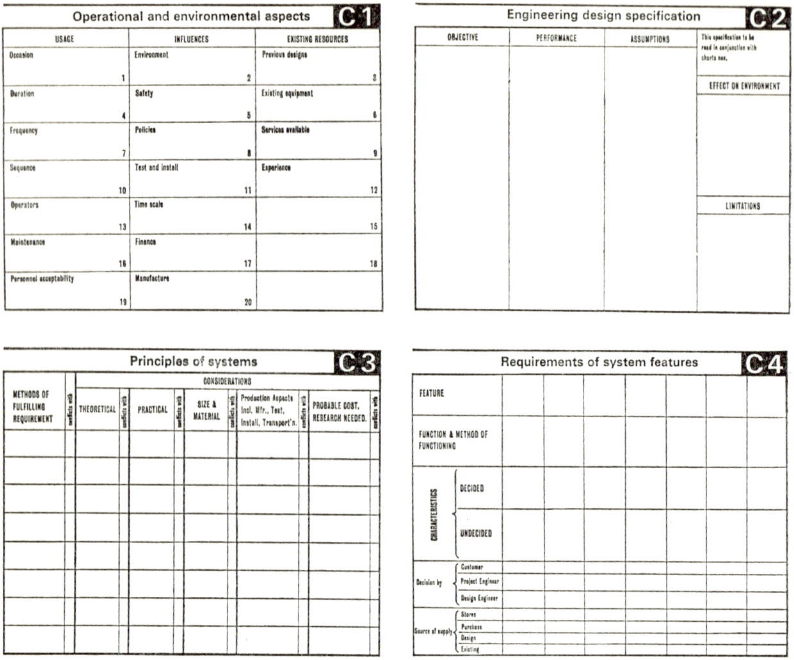

Figure 2.3 The four PABLA charts which are completed and evaluated, as a part of the PABLA system, before drawing can be started

tion to search for established methods he may be unnecessarily preconditioning his thoughts towards stereotyped design.

C2 continues the specification by encouraging the user to consider the basic purpose under the heading of objective. The performance column provides an opportunity to list the effects that the various items, listed on C1, have on the objective. The heading 'assumptions' speaks for itself. An unusual and useful feature of the PABLA method is a consideration of the design's influence on the environment. The limitations section is intended for constraints resulting from safety and economic considerations, etc.

The facility for the generating of alternatives is the first column of C3. The only stimulus provided, however, is the need to fill the empty spaces provided. In fact, it may be necessary to resort to other techniques such as brainstorming in order to complete the column. The remaining columns are devoted to the evaluation of the alternatives, the first two on technical grounds and the remainder on economic bases. C3 is a good example of the importance of superficially considering the various aspects for the purposes of assessing alternatives. The reason being that when the form is completed, a scan will indicate the promising solutions, i.e. technical and economic assessments are in effect being made simultaneously.

Finally, C4 provides the necessary data for synthesis enabling the designer to proceed to the stage of communication, or drawing. A major attribute of C4 is that its layout makes the user aware of new design requirements originating out of the solution to the original problem.

2.9 THE MORPHOLOGICAL METHOD

This approach is attributed to Zwicky[7] and its use in a commercial design office is described by Norris[8]. The method hinges around a morphological chart or design matrix (table) and really only caters for the stages of presenting and evaluating the alternative ideas. To use the matrix for this purpose it is of prime importance that the designer has carefully established his specification; to employ the morphological matrix without first doing so could result in obvious chaos. The morphological approach derives its name from the dictionary definition: 'pertaining to the study of an organised system or form'.

The matrix comprises a single left-hand column in which are listed the parameters essential to the design, i.e. what the design must be, or must have. To the right of each element in the column is a row containing the possible ways of achieving that particular parameter. This contrasts with chart C4 of the PABLA system where only a single requirement is studied per sheet, emphasising the fact that the morphological method is principally concerned with the speculation step in the process.

To illustrate the use of the morphological matrix, consider the design of a forklift truck for use in a warehouse. The design parameters resulting from the specification would include:

(i) means of support which allow movement across the floor
(ii) means of steering the vehicle
(iii) means of stopping
(iv) propulsion
(v) power unit
(vi) power transmission
(vii) lifting mechanism
(viii) facilities for operator

A morphological chart showing these parameters and some possible ways of satisfying them is shown in Table 2.2.

If every solution on each row is compatible with all the solutions on the other rows the number of possible solutions to the system is

Table 2.2 A MORPHOLOGICAL CHART FOR THE MAIN REQUIREMENTS OF A FORKLIFT TRUCK

Parameters	*Possible solutions*				
Support	Wheels	Air cushion	Tracks	Slides	Spheres
Steering	Turning wheels	Rails	Air thrust		
Stopping	Reverse power	Brakes	Blocks under wheels	Drag a weight on the floor	
Moving	Air thrust	Power to wheels	Hauling along a cable	Linear induction motor	
Power	Electric	Bottled gas	Petrol	Diesel	Steam
Transmission	Hydraulic	Gears and shafts	Belts or chains	Flexible cable	
Lifting	Screw	Hydraulic ram	Rack and pinion	Chain or rope hoist	
Operator	Seated at front	Seated at rear	Standing	Walking	Remote control

a multiple of all the ideas on the rows, and even for the small field given in *Table* 2.2 the possibilities would be enormous. Clearly many of the alternatives are not compatible, e.g. an air cushion vehicle could not be steered by its wheels if it had none.

In the example of the forklift truck some of the alternatives along a row may be combined to give a single solution, e.g. hydraulic transmission may be used for the lifting mechanism whereas the drive to the wheels might be in the form of shafts and gears. It can be seen that transmission may have been divided into lift and motive; any morphological chart may be subdivided to cover all aspects of the problem. The coarseness of the division is decided by experience. This example illustrates the problems which may arise if the specification stage is ignored. By formulating a specification it would have been shown that steering by rails is unacceptable for a general purpose forklift vehicle. Similar arguments can be applied to some of the other solutions suggested.

In the example of the forklift truck the truck was required to perform various functions, which were listed as design parameters. An alternative is to consider the properties required of the design. Consider a tool for tilling the soil; it will not be described as a spade as this suggests a particular solution to the problem. The parameters may now be divided into functions and properties. The only function is that it should employ manual effort and result in the ground being broken up:

(i) breaking up the ground

The properties required are:

(ii) light to use
(iii) ease of stowage
(iv) corrosion resistance
(v) simple to maintain
etc.

Each of these parameters can be listed on a morphological chart as in *Table* 2.3.

Table 2.3 MORPHOLOGICAL CHART FOR A TILLING DEVICE

Parameter	*Possible solutions*			
Break ground	Dig	Plough	Explode	Rotate
Light	Use light material	Make of slender construction	Fill with helium	
Stowage	Make small	Make flat	Design to fold	
Corrosion resistance	Protective coating	Use corrosion-resistant material	Allow to corrode over a period of years	
Maintenance	Use simple design	Add devices to assist maintenance, e.g. centralised lubrication system		

2.10 VALUE ANALYSIS

Value analysis (VA) is essentially a redesign method, although the associated discipline of value engineering involves original design. VA is a team activity, the idea being that an item already in production is critically examined for value, with a view to redesigning it for greater value. Value is interpreted in the widest possible sense to include not only the monetary cost of the component, but also its ability to function satisfactorily. The VA team is drawn from many different fields, e.g. production, design, sales, research, subcontractors, etc., so that a wealth of experience may be brought to bear on redesigning the product. Value engineering takes this approach at the design stage. Here the team studies a new design, rather than an existing product, and advises the designer on possible improvements before the product proceeds to manufacture. This is not strictly a design method in that a design must exist in order for the team to be able to evaluate it. However, in the United States, Davis[9] states that the term 'value engineering' has come to include all studies of design method.

Value analysis is of interest in that its stages of operation can be studied in terms of the design process. These stages have been given various names. The descriptions listed below are in fairly common use. These steps are:

(i) Information
(ii) Speculation
(iii) Investigation
(iv) Recommendation
(v) Implementation

The information stage requires the gathering of factual information about the component under the direction of the team leader. This is representative of the specification in the design process. Basic questions have to be answered, e.g. what function must the component perform, is it really required?

Speculation is the search for alternative ways of meeting the specification and, since a team is involved, brainstorming provides the usual vehicle for generating these ideas. Investigation is concerned with the feasibility, both economic and technical, of the various solutions, and is directly analogous to design evaluation. Recommendation is the passing on of the solution from the value analysis team to those concerned in the redesign, and as such has no equivalent in the design process. Implementation is directly equivalent to synthesis.

Value analysis is a technique adopted by many industrial organisations and serves to illustrate one way in which the design process is effectively coping with the day-to-day problems of industry.

2.11 EXERCISES

EXAMPLE 1

Requirement (from customer)

A pair of buffers is required to stop small trucks used for mining minerals. The buffers are located at the bottom of the mine, near the cage which hoists the trucks to the surface. The trucks have been slowed down to a maximum speed of 20 m/min at the time they strike the buffer, and the maximum laden weight of each truck is 2000 kg. Fixing is to a flat vertical face. The buffer must return to its original position after impact and there must be no recoil.

Specification (Compiled by Designer)

The function of a buffer is to dissipate energy. Kinetic energy to be dissipated per buffer $= \frac{1}{2}mv^2$

$$= \tfrac{1}{2} \times 1000 \times (\tfrac{20}{60})^2 = 55{\cdot}5 \text{ J}$$

If the stopping force is to be a minimum, and since the area under the force–displacement curve is equivalent to the energy, it will be seen from Figure 2.4 that the force should be constant over the stroke.

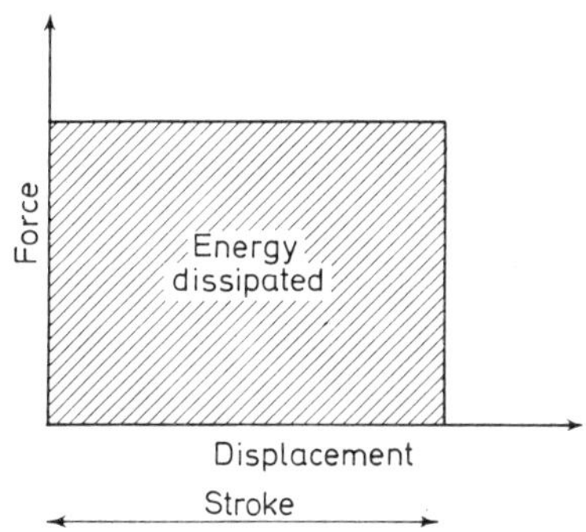

Figure 2.4 The ideal force–displacement characteristic for a buffer

Although a space restriction has not been made, bearing in mind other possible applications it would be unrealistic to have a buffer longer than 0·5 m with a stroke of say 0·25 m; this results in a constant retarding force of 222 N. The last sentence of the requirement is vague. No recoil means that it must return relatively slowly to its original position, it must also return in time to stop the next truck. A time of between 10 and 60 seconds would seem appropriate.

The unit must also: be protected from contamination by dust, be able to tolerate slight misalignments of the truck, have suitable provision for fitting, require the minimum of maintenance (preferably none), not present a fire hazard, be easy to pack and transport, be competitively priced.

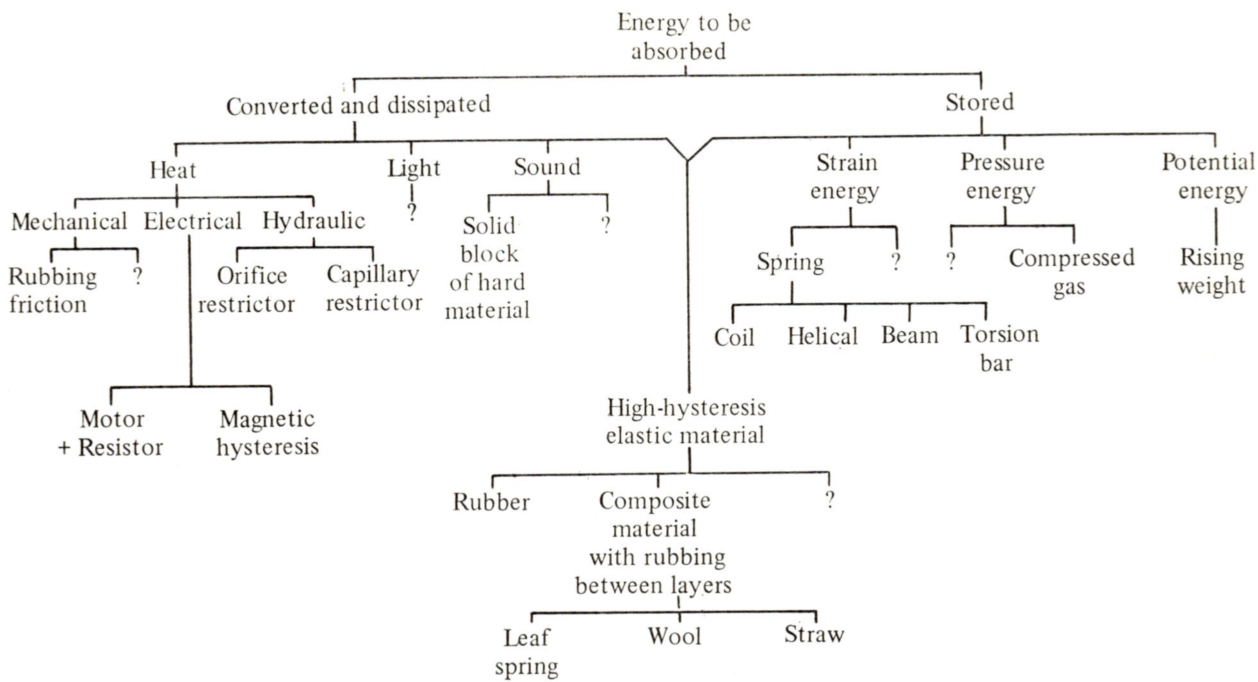

Figure 2.5 The ways in which a buffer may absorb energy

These are all essential features of the product; the reader may be able to add to this list. Note that unnecessary restrictions, e.g. method of operation, colour, shape, etc., have not been included.

Speculation

There are two ways in which energy may be absorbed: it may be stored as strain energy or pressure energy, to be released later, or it may be converted into heat or sound and dissipated. It will be seen that commercially available buffers make use of both forms.

The possibilities are laid out as a family tree in Figure 2.5.

Evaluation

The following notes, though not exhaustive, are a guide to evaluation.

(i) Selection must be related to the energy being absorbed and the space available, e.g. the buffer for a light instrument slideway could be made from wool felt, or a racing car could be stopped by bales of straw.

(ii) It is unlikely that any of the suggested solutions in their simplest form would produce the desirable characteristic shown in Figure 2.4. They must either be combined with other solutions or be selfcompensating in some way.

(iii) Although it would be easy to dissipate the small amount of energy electrically, through a resistor, it is doubtful whether a magnetic coupling of sufficient strength could be created within a suitable working clearance. There may also be a fire hazard.

Synthesis

A commercially available buffer will be described to show how the various principles are embodied in practice. This does not imply that this is the only satisfactory solution to the problem. A sectional diagram of the buffer is shown in Figure 2.6a. It can be seen that a composite system is employed, part of the energy being stored by compressed air, the remainder being dissipated as heat in the viscous flow of oil through the orifice. The stored energy repositions the buffer after the load has been removed. Figure 2.6b shows the force–displacement characteristic of the air chamber and Figure 2.6c the characteristics of a buffer using hydraulic oil flowing through a fixed orifice. In the buffer shown the loss characteristic is modified by using the tapered metering pin. The combined result is the characteristic shown in Figure 2.6(d).

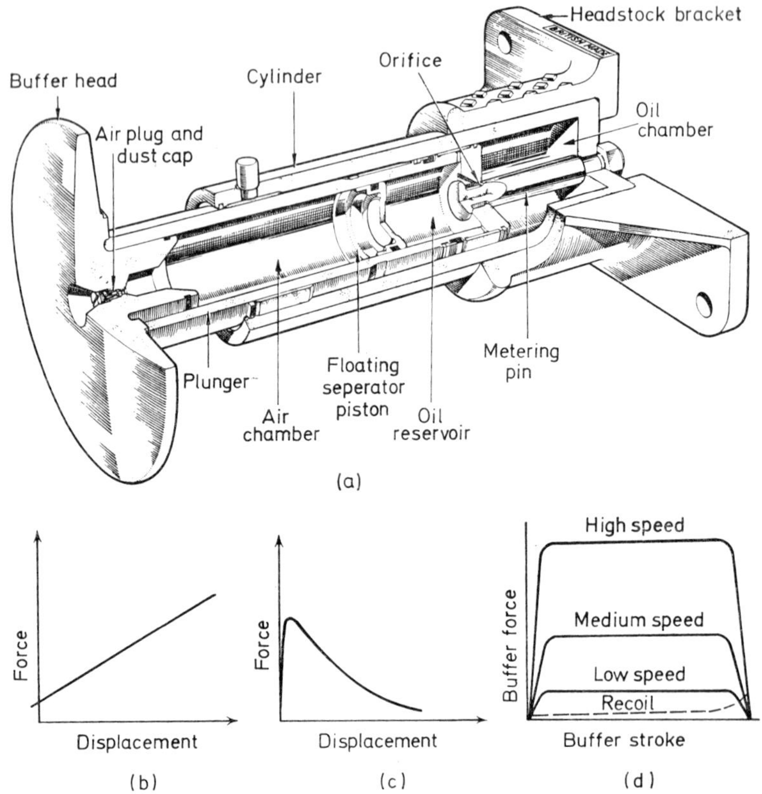

Figure 2.6 The buffer shown in (a), produced by Oleo Pneumatics Limited, comprises a compressed air chamber and a hydraulic orifice controlled by a tapered metering pin. This gives the characteristic shown in (d). (b) is the characteristic of the compressed air and (c) that of a simple hydraulic buffer having a fixed orifice diameter.

QUESTION 1

Repeat Example 1 using the PABLA and Morphological methods. Make notes on the limitations of each method and any advantages they have.

QUESTION 2

A need exists for a contents gauge to be installed in a water storage tank. The tanks are usually 20 metres and 40 metres in diameter and between 20 and 40 metres high. The customer has asked for a gauge capable of measuring the depth to within 5 mm.

The last statement is irrelevent to the specification, what the customer really needs is a gauge accurate to within a certain volume, this accuracy can be calculated on the basis of the smallest diameter tank. The market from the requirement is limited, and could be enlarged with a gauge suitable for petroleum storage; here a fire hazard exists, or for chemical storage if the gauge is made of materials compatible with the chemicals.

Repeat the exercise outlined in Example 1. Start by considering the parameters which change as the tank is filled, e.g. liquid height, hydrostatic pressure. Consider also the possibilities of metering the difference between inflow and outflow. Continue the process by listing mechanical, electrical, acoustic and hydraulic ways in which these might be measured.

A wide range of these gauges is on the market and the reader is certain to find that many of his ideas are already implemented.

QUESTION 3

Garage doors are of many forms, and within any given type there may be several ways of creating the desired movement for manual operation. Automatically opening doors are becoming more common, but many manufacturers add the automatic mechanism to the manual door as an afterthought.

Methodically list the different types of door together with the ways in which movement is made possible, and then list probable drive systems for the doors. Assume that a normal mains electrical supply is available. Make notes on the compatibility between the systems.

QUESTION 4

Domestic off-peak electric storage heaters take in a charge of electricity at night to heat the core and then give off the heat the following day. Adjustment to the heaters has to be made in the evening, or during the night for the following day. An automatic adjusting device would have to sense the night-time weather conditions to relate the following day's temperature to the heat requirements of the building, and hence adjust the charge of electricity supplied to the heater.

Investigate the relationship between the outside temperature and the heat requirements of the building, taking account of wind and rain etc. (The relationship should prove to be almost linear.)

Establish a family tree of devices capable of sensing the external weather conditions and relating them to an electrical charge. There are several parameters which give a measure of weather conditions; bear in mind that this is a forecasting device. Equipment designed for this purpose is already marketed. Find out its method of operation and see where it fits into the range of possibilities considered.

QUESTION 5

Shops trading in unwrapped food, such as meat and dairy produce, must, according to UK law, separate the handling of the food from the handling of money. Use the design process to devise a machine which will receive payment from the customer and enable the salesman to give change without touching the money. The machine must be acceptable and nonconfusing to the customer, and should be capable of dealing with different currencies, including paper money.

3 Economic implications of design

3.0 ECONOMY OF EFFORT

The designer should not fail to recognise the pound, dollar, or franc, as an engineering unit as significant as the kilogram or the metre. In spite of the development of the economic sciences and the introduction of improved costing techniques, the subject still remains somewhat nebulous when compared with the more predictable calculations of applied mechanics. This point is emphasised by the reluctance of component manufacturers to provide any indication of price as a part of their product data.

In order to understand manufacturing economics, it is essential to realise that cost is a measure of human effort. On a wider basis it may be wrongly argued that this thesis does not apply to the provision of raw materials. Minerals, for example, may rise in price owing to the difficulties in mining them or owing to their general scarcity. Scarcity is still related to human effort, a good example being diamond; a shortage of industrial diamonds would increase the complexities of machining hard materials; in a similar way an abundance of platinum would have made unnecessary many of the developments in nickel–chrome steels.

The underlying principle of economic manufacture is economy of effort. Even the builders of ancient churches and palaces were aware of this philosophy. The prime purpose of such buildings was to be beautiful rather than functional. Although many of the pillars appear to be carefully hewn and carved, this quality is only skin deep. The economy of effort manifests itself at the centre of columns, where irregular stones are cemented together in a random fashion. Unfortunately, many of these early builders were so intent on placing the effort where it could be seen that little or no attention was paid to the construction of proper foundations and many of these beautiful buildings have now become a liability to their owners. Probably the best example in this field is the Taj Mahal at Agra in India. The Taj is finished in white marble, inset with semi-precious stones, but the basic structural material is a common red stone of the area. The Moguls conquered Agra and built the Taj; the British arrived and added a lightning conductor; India's own government is now considering ways of reinforcing the already inadequate foundations. The designers of the Taj even overlooked a visual flaw which lays bare their attempts at economy. On the roof of the Taj, at either side of the great dome, are two walls. These walls are finished with marble on their outer facing sides and at their ends, but on the sides facing the dome the basic red stone remains unfinished. The effect on an elevational drawing remains unnoticed, but perspective enables the visitor, viewing the Taj from the front, to see this blemish.

3.1 THE PLACE OF ECONOMY IN ENGINEERING DESIGN

In design one is normally concerned with relative costs; usually a specific requirement has to be fulfilled and the designer is only concerned with meeting this requirement in the most economical way, though an aspect such as sales appeal may warrant some expenditure on aesthetics. Problem recognition, and the ability to analyse the requirements for the specification, are especially important in design economics. For example, there is little to be gained in designing a component within a system for a greater life than the system itself, or designing a component in a measuring device to meet a system accuracy of 1% when the overall accuracy is only 15%. In other words the designer should be asking himself 'What is the minimum satisfactory requirement?'. On a broader field, the designer should be aware of the effect his design might have on the final cost of the product. Whilst it is useful to have at hand the absolute or relative costs of the various manufacturing processes, it is far more important for the designer to be aware of the areas where expenditure may be incurred in order that he may design for economy. Actual costs soon become invalid, and many of the relative costs may change with time. For example, a decade or so ago, point-to-point numerical control was considered uneconomic by most industries, whereas currently it is one of the most efficient ways of handling many of the batch production machining operations.

Figure 3.1 shows diagrammatically some of the costs which have to be borne by the price of the finished product. The contributions made to the whole by the various subdivisions will depend largely on the particular product being manufactured and the production throughput. It will be seen that a deliberate degree of duplication occurs on

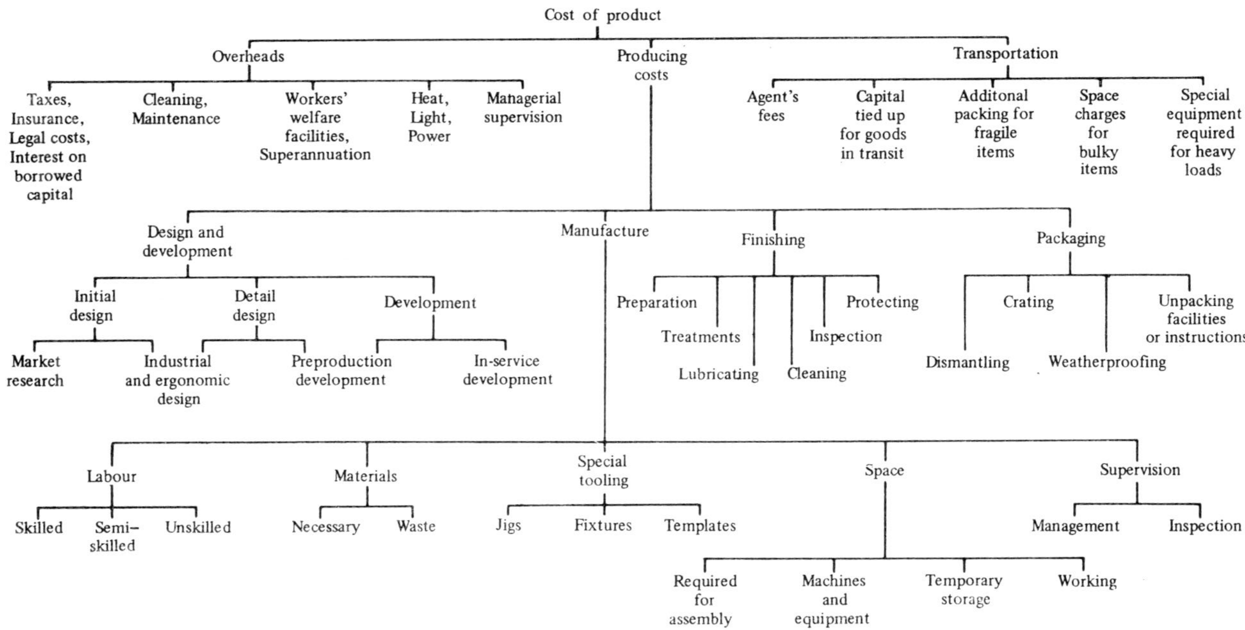

Figure 3.1 The breakdown of product costs

this diagram, e.g. managerial supervision appears as a subdivision of manufacture and as an item of overhead expenditure; the reason for this duplication will become apparent later.

The designer too frequently considers that the only contributions made towards the product cost are those resulting directly from manufacture, and fails to realise the financial implications that his design may have on the other areas of expenditure. In an attempt to redress the balance, the following discussion will lay a particular emphasis on those areas outside manufacture.

3.2 OVERHEADS

In some situations it is difficult to define exactly what is, and what is not, an overhead. This point is best illustrated by considering the two extreme conditions of a one-off item passing through a general jobbing workshop, and the massproduction of a particular model of motor car at a specially built factory. In the latter case it is possible to identify a certain grade of supervisory management with the manufacture of a particular item or subassembly; here the degree of supervision directly relates to the price of that item. On the other hand, the one-off product shares the general supervision of the workshop with any other items which may be passing through during the same period, and hence the only method of costing is to make a blanket percentage addition to the product price; the blanket charge then becomes an overhead charge on the product. Although it may be possible to identify a supervisory duty with an item, some companies prefer to carry the supervision as an overhead.

From the design point of view, the essential property of an overhead is that it appears as a percentage surcharge on the item being manufactured. Most of the expenses contributing to this surcharge are shown in Figure 3.1.

At first sight it would appear that overheads are completely outside the control of the designer. Whilst this is true, the designer can still affect overheads; with good management in control of the situation, design economies can be exploited. But how can the designer reduce overheads? First of all an overhead is normally a percentage surcharge based upon the manufacturing cost of the product. By reducing the basic product cost a related reduction in surcharge results. Secondly, a designer having a sound knowledge of his companies' facilities and the way in which they are financed can make further savings. A good example of such a situation is workshop loading. When faced with two alternatives to a particular problem, and one requires the services of a heavily pressed section of the manufacturing plant, and the other would take up the slack in some other section of the works, then clearly the latter is to be preferred. If, however, the opposite decision was taken, extra machinery and personnel may be required to meet the demand, leaving the underworked department as a liability upon the company.

A further example of good design cutting company overheads is in the field of rationalisation. By reducing the variety of items produced, the capital and space invested in spare parts is also reduced; variety reduction in raw materials leads to better purchasing and as a consequence ready availability.

The economies already mentioned suggest a curtailment of factory services and possibly the redundancy of employees, so how does good management allow the overhead per item to be reduced and at the same time find cash to finance these services? There are basically two ways: either increase production if the market is not already at capacity, or diversify the manufactured products. The latter has the advantage of making the company less vulnerable to market fluctuations.

Summarising:

(i) Overheads can be reduced by simplification of the product and by designing to use spare production capacity.

3.3 TRANSPORTATION

Transportation of the finished product is an area in which poor design can wreck commercial prospects, particularly on an overseas market. An important point to be considered is whether or not the manufacturer is responsible in some degree for product delivery. In a majority of cases the manufacturer is expected to bear the cost of delivery of the goods. Exceptions to this policy might include ships or aircraft, where the customer undertakes to collect the craft from its constructor; here, delivery cost has little effect on the market price.

Unfortunately, most designers only want to know whether the product can be moved from A to B. The shipper, with his professional prowess, will agree to move almost anything, but unless requested fails to reveal his fees for the service. A well designed product simplifies transportation.

Most designers have the impression that shipping charges operate on simple linear scales; they fail to realise that scales often rise in steps, or have an upper limit. The Appendix provides a guide to the reader. It will be seen that it is very easy to incur a heavy financial penalty by designing an article which slightly infringes a maximum standard dimension, e.g. a few millimetres on a road-borne item could make the difference between needing a police escort or having a routine shipment. To use container facilities efficiently, it is essential that the total weight of the contents is near to the maximum payload of the container. If a single item three quarters fills a container, could it have been designed for a smaller enclosure or for packing with similar items to provide a full load?

The structural suitability of a product to the mode of transportation is also important; packing costs money and this money can be saved or be better spent on the product itself. In addition to the simple impact loads which are encountered in transit, there are more subtle risks such as environment, e.g. sea spray, or the problems of decompression encountered in air freight. When fragile items are being shipped the liability normally rests with the owner; it is essential therefore that the articles are packed in such a way that their container may be carelessly handled without prejudicing the safety of the contents.

Frequently, the whole assembly has to be over-protected for one minor component; by designing for easy removal of fragile components, which may be packed separately to a higher standard, the remainder can be protected in a less demanding way. One such example is the removal of valves (tubes) from large electronic devices. A further example of good design is provision for anchoring 'floating' components during transit.

It can be seen from the Appendix that standard packing densities exist. A penalty is paid for transporting bulky items, i.e. a density less than that stated. Here a charge is made for the equivalent volume, this may mean that the customer is paying for the transportation of trapped air. The density limit varies according to the mode of transport, but one common fact emerges; a good design requires high packing densities; two obvious examples are: stacking furniture, and the turning of bicycle handlebars and pedals prior to shipment.

Although bulk is usually a greater problem than weight, sometimes unnecessary weight is moved, particularly where the weight is required as balast by the recipient. Balast can often be provided more cheaply nearer to the final siting. Concrete blocks are used for counterbalancing civil engineering contract equipment, and sand or water may provide filling for surface rollers. When a product has to be transported over great distances the filling of lubrication or coolant reservoirs at the destination may reduce costs substantially. The elimination of unnecessary weight is particularly important with air freight. The attractiveness of air freight from the manufacturer's point of view is twofold: better delivery times and, as a consequence, earlier payment for the goods.

Summarising:

(i) Freight charges do not increase at a simple linear rate.
(ii) High packing densities and low overall weight reduce transportation costs.

3.4 DESIGN AND DEVELOPMENT

The small expenditure required to operate an effective design office, when compared with other sections of industry, has led to the design office being considered something of a poor relation, which in the limit results in a fall of product standards. In those industries where design expenditure is a small proportion of the total, the quality of design becomes even more significant. These are industries with a high production capacity where poor design at the outset can spell economic disaster. Many such companies depend on a heavy development programme to right errors which could have been detected at the design stage.

Good design costs money and 'perfect' design requires unlimited financial resources and time. It is the designer's responsibility to decide the amount of time that he can effectively afford to devote to a problem. His decision must take account of the final market price of the product, the numbers to be produced and the hourly rate for his own services and the design office overheads. Consider a simple bracket required to support a small valve in a chemical plant. If only one bracket is required and it happens to be a nonstandard part, then almost any bracket satisfying the technical requirement would be acceptable, in which case the designer may sketch out the first idea which enters his head. To spend time carrying through the full design process could produce design costs several orders of magnitude greater than those of manufacture. In contrast, the design of a simple bracket for a popular motor car could justify several man days of design work. Here the saving of a few pennies on each bracket would, because of the quantities manufactured, more than offset the expenditure in design.

Whilst design is normally governed by economic considerations, exceptions do exist; here the designer finds himself under a moral obligation to the community. Such an example is in the aerospace industry where defective design can result in loss of life. Even if the designer fails to recognise his moral obligations he should still detect such situations from an economic viewpoint, i.e. failure resulting in loss of life and extensive damage are costly, prevention being cheaper than cure.

The allocation of expenditure on design is complex. The two general areas in any design activity are the initial design and the detail design. This is an oversimplification in that a grey area exists between the two, but it serves to illustrate the way in which costs are allocated. Initial design can be financed in one of two ways: either the company is prepared to speculate on a new product, in which case the cost of design will be met from the company overheads, or the customer commissions the work and makes a payment in return for the service. In the former situation the design activity may well be linked with some form of market survey, often carried out by an independent body; the triangle may be completed by the involvement of industrial design consultants to help give the product an image for purposes of sales appeal.

By the time the detail design stage has been reached the project is normally well established. Although some industries prefer to consider all design activities as overheads, many firms, particularly those supplying the batch markets, cost the detail design against the product itself. It is here that the designer must carefully cost his own efforts against the savings which may be made in simplified production.

Development can take place at any stage in design and manufacture. At the predesign stage, a research project may be undertaken to investigate a basic physical phenomenon with a view to exploiting it in some way. When a design has been completed a number of doubtful quantities often exist, e.g. a structure may have been designed using the latest analysis, but in some areas approximations have had to be made and require physical tests for their justification. During production it is sometimes necessary to develop a design to facilitate easier manufacture. Finally, product shortcomings may be discovered during service and some work will be necessary for their rectification.

Development costs are to a large extent related to the problems of communication. Where development takes place in the design and development complex, communication is simple and the integrated effort of the design and development engineers is likely to be highly effective. At the other end of the spectrum, a product requiring design modification, resulting from a failure during service overseas, involves expensive and less effective communication which at the same time is slow. Here verbal contact is difficult and ideas have to be exchanged through the medium of written correspondence resulting in several days elapsing before agreement can be reached on the nature of the problem and the modifications necessary to correct it. Another, and very expensive, aspect to post-design development is the scrapping of manufactured components. This is more serious in massproduction, where an army of service engineers has to be recruited to trace and rectify those products already in use. The point is brought home when motor manufacturers find it necessary, for safety, to replace a particular component on all the cars they have manufactured. This may cost several million pounds, whereas a little more expenditure at the design and preproduction development stages would have prevented this arising.

Summarising:

(i) The designer must compare his efforts with those of manufacture and consider any moral implications of his work.

(ii) In general, preproduction development is less expensive than that carried out during service.

3.5 FINISHING

It would seem logical to consider the subject of finish at the end of a discussion on manufacture. The word 'finishing' suggests the final operation of the manufacturing process, and for this reason many uneconomic designs result; the finishing operation can often be completed as an integral part of a process part way through manufacture.

In some instances the cost of finishing a product may exceed the cost of manufacture. A good example is that of moulding. Often, production quantities are such that a simple sand casting is marginally cheaper than an equivalent die casting, but when it is considered that the sand casting may have a flash, runner and riser, which must be removed, and that it must be painted to improve its general appearance, it may prove cheaper to employ an unpainted die casting in the condition it leaves the die casting machine. An even better example of forming is the use of moulded plastics; in addition to a high standard of surface finish, the item emerges ready coloured. The steel used to make a standard domestic bucket is very much cheaper than the equivalent plastic, but since it is necessary to galvanise or enamel the metal, the plastic bucket proves to be less expensive.

The finishing operations in manufacture are preparation and treatment. Many designers, in specifying a particular treatment, fail to recognise the degree of preparation necessary before the treatment can be implemented. Generally, the preparation can be more costly than the treatment itself. All forms of preparation involve 'cleaning', and many manufacturing processes result in a 'clean' component only to be 'dirtied' before treatment; so why not carry out the treatment at this point in manufacture? The only criterion that has to be observed is that any subsequent processes do not damage or destroy the treatment. An example is the spotwelding of aluminium: the surfaces have to be cleaned in order to obtain a satisfactory bond; to follow this by painting or anodising has clear economic advantages.

In specifying treatment it is necessary to study the exact requirements which have to be met. The function of a treatment is of primary importance; is it to protect, to give an appearance, to improve the surface wearing properties, or to provide a certain surface texture, etc.? Sometimes the designer spends several hours in evaluation, and then gives little or no consideration to treatment. The number of treatments available for metals alone is large, and divides roughly into two groups: those which are intended to alter the physical properties of the materials, such as hardening and tempering, and those which are intended to form a surface layer for protection or appearance. Some treatments, such as the anodising of aluminium alloys, result in both. It is in surface treatments that the greatest variety arises.

In selecting a surface treatment, from the economic aspect, the following points should be considered:

(a) By using a different material can the treatment be avoided and the overall component price reduced?
(b) What treatment facilities exist within the company or in the immediate geographical area?
(c) Does the quantity involved warrant the setting up of special equipment?
(d) Can the treatment selected be integrated into the manufacturing process?

In addition to the points already made, it is important that the designer is aware of the wide range of treatments available to him. Lubrication, cleaning, inspection and protection can also be integrated into the manufacturing sequence. The application of lubricants to the finished product can be costly and time consuming. Although a prelubricated or selflubricating bearing may be more expensive than its dry counterpart, the comparison for cost must include the addition of the lubricant; not only is this a further task to be performed, it may add substantially to the final cleaning.

It should be clear that, where cleaning is a consequence of manufacture, it is often more economical to maintain the state of cleanliness than to start all over again with the cleaning operation. The designer can cater for this by employing sealed units where cleanliness is essential. These units can be manufactured and sealed in a controlled environment and then added to the final assembly. This technique is employed in the manufacture of domestic refrigerators, where the compressor is supplied as a sealed unit; such designs facilitate easy replacement during service. An undesirable side effect to the manual cleaning of items at the end of manufacture is the damage which may be caused. If cleaning must be carried out before the product leaves the factory, then good design will result in easy access, or the design will lend itself to an automated cleaning process. Trapped debris leads to unreliable products and can prove fatal in aircraft manufacture, etc.

Summarising:

(i) Can finishing be made an integral part of the manufacture?
(ii) Treatments are very varied, but a properly selected treatment can enhance and cheapen a design.
(iii) A good design will provide easy access for cleaning and inspection.

3.6 PACKAGING

The price of a product does not end with the product itself, but must include the price to be paid for its 'wrapping'. The purpose of packaging is, in most cases, to protect the product; the exception of course being display packaging for sales promotion. Protection can be broken down into two general categories: sealing, required to protect the item from the chemical effects of the environment, and 'cushioning' for protection against shock loads.

Protection against the chemical effects of the environment may be achieved in one of two ways: either the packaging can be made in such a way that the item is sealed from the hostile environment, or the item may be designed to withstand the environment without any need for protection. Clearly the latter arrangement would seem the most desirable, since the product would have protection throughout its working life. Such a design decision can only be based on product economics: which is the cheaper of the two alternatives? If a negligible cost difference exists, then naturally the permanently protected product is to be preferred. Fortunately for the majority of designers the sealing of products is a relatively cheap operation and in most cases the fact that this has been overlooked has a negligible effect on the final price.

For the protection of products against shock loading, a similar argument can be made to that of sealing. The greatest design mistake made in this area is the failure to realise that an impact on the side of a crate is not just simply a rapidly applied force, but it imposes high accelerations on the components within. The function of the package is to prevent direct damage from the impact and to dissipate the impact energy before it is transmitted to the contents. If inaccessible components are designed in such a way that they are unaffected by high accelerations, then packing is reduced to a simple external protection. Where a high degree of packing is required, this can be carried out more cheaply if consideration has been given to the physical shape of the product: large overhangs not only require extra space, they also require support; sharp corners puncture plastic sealing films, and inaccessible hollows prevent sprayed sealants from reaching the whole surface.

Summarising:

(i) Can a part, or all, of the packing be eliminated if the money is spent on improving the quality of the product?
(ii) Designs should make for ease of packing.

3.7 LABOUR

All manufacturing processes, no matter how automated they are, require some degree of human intervention. To divide labour into the three arbitrary categories of skilled, semiskilled and unskilled might at first sight seem illogical, but since the cost of labour, or the cost of avoiding its use, may account for the greatest proportion of a product's price, the designer should be aware of its abilities and limitations. It may be argued by some that a semiskilled operator assembling expensive electronic, or other components, contributes little to the price of the finished product, but what they fail to realise is that each item may already incorporate a large proportion of manual effort.

How can the designer influence labour costs? Since different grades of skill exist, somewhere along the line between the designer and the maker someone has to decide upon the degree of skill required to complete the work; this task usually falls upon the production engineer. If the job is to be carried out by an unskilled employee, then either it must be extremely simple, possibly trivial, or alternatively special tools must be made to aid the operator. Clearly, for complex jobs there is no alternative to using the skills of a craftsman. The traditional belief, that unskilled labour is relatively cheap, and skilled labour is very expensive, is not always valid. Although there may be a nominal differential between the two rates of pay, in fact the unskilled worker is often well paid to compensate him for carrying out a monotonous and sometimes trivial job. Furthermore, the unskilled worker normally requires a greater degree of supervision and special tools or machines for the purpose. In deciding the method of production, the production engineer is more likely to be influenced by the numbers required, if the number is large, special tools may be called for, which could in turn lead to the use of unskilled and semiskilled manpower. On the other hand, the cost of special tools cannot be justified where quantities are small. The designer must take account of this fact, particularly at the detail design stage. If he is aware that quantities will be large, he can ensure that the component parts are suited to manufacture by automatic machine tools, and that assembly can be automated or carried out as a series of simple operations by unskilled operators. He must also bear in mind that in these circumstances it is unlikely that the drawings he produces will be used by the individuals employed in manufacture, but will provide the information for the tool designers and the production engineers preparing the operator's instruction schedules.

For small quantities there is usually no alternative to the use of skilled labour. To gain the maximum benefit from these skills the designer should avoid the use of special tools and place some reliance on the craftsman's judgement. For example, consider a complicated casting, the contours of which are all curves and into which a bearing housing is bored. For quantity production a special fixture would be made to locate the boring bar relative to the casting contours. If the quantities are small the setting up would have to be carried out by hand, and the use of a special fixture could not be justified; the designer would then provide flat pads on the casting for purposes of location. All too frequently these needs are overlooked, and design modifications have to be requested by the production engineer.

An important aspect, in the efficient use of labour, is the system of incentives. These may operate in one of two ways: either block incentives, where a group of employees share the financial benefits of higher productivity, i.e. profit-sharing schemes and group bonus systems, or individual incentives, i.e. piecework payment and individual bonus systems. In general, group schemes appear to work most satisfactorily with groups of about a dozen, with larger groups much of the competition and enterprise is lost. The normal method for assessing bonus is for each task to have a fair time estimated. If the employee completes the work in a shorter period, then he is paid a bonus for the job, being the difference between the estimated and actual time. The piecework system, mainly operated at the unskilled and semiskilled level, involves a specific payment for the completion of a particular task; here again an estimate has to be made to establish a fair price.

How does the incentive system affect the design? To establish a fair time for the job the designer's drawings are studied. Every instruction on the drawing takes time to carry out; the more instructions the drawing conveys, the longer the time required, and consequently the greater the price to be paid. Problems can arise in bonus agreements if designers insist on the liberal use of tolerances even though most of these may be quite large; a superficial examination by the operator or ratefixer might suggest a precision job. An example of this is shown in Figure 3.2. The tolerances shown in Figure 3.2(a) have been incorporated in a blanket machining tolerance of $\pm 0{\cdot}5$ mm in (b); as a consequence the two critical dimensions become clearly visible. Even though some of the tolerances in (b) are now tighter then those in (a) it is likely that the work would be costed at a lower-value. Deviation from adopted standards can also cause difficulties in rating a job.

Summarising:

(i) The skill of labour is more often related to the quantities required.

(ii) Every line on a drawing requires time and effort to interpret and implement.

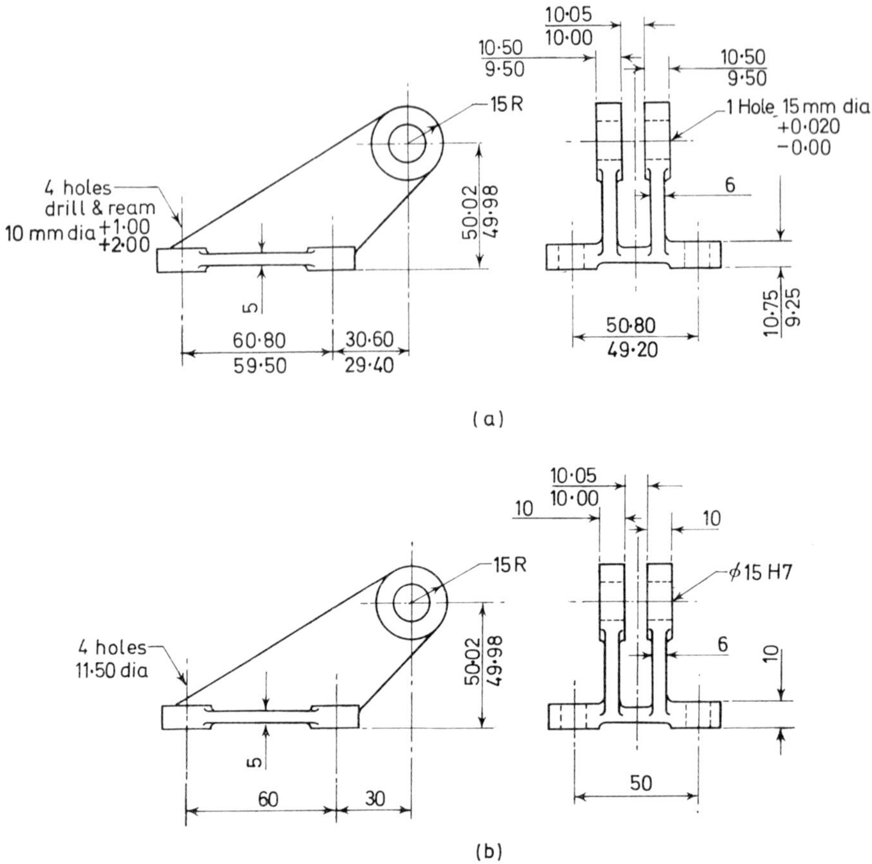

Figure 3.2 The visual effect of an over-toleranced drawing makes it difficult to recognise the critical dimensions and leads to difficulties in costing the precision of the job. In (a), most of the tolerances could be covered by a general machining tolerance of $\pm 0{\cdot}50$ mm *which would be cheap and simple to achieve, as shown in (b)*

3.8 MATERIALS AND WASTE

Materials in this context are taken to include raw materials and components, i.e. nuts, bolts, fasteners, etc. The same basic rules apply to both raw materials and finished items. The designer is primarily concerned with the material costs relative to other costs such as labour and overheads. The desirability or otherwise of having a large contribution made to the final product price by the material from which it is made, depends largely upon the economic environment in which the designer works. If the environment is one having large human resources where materials have to be imported, an extreme example being Hong Kong, then the optimum design is the one in which the bulk of the material is in the raw form and, consequently, the final product price is influenced to a greater degree by the cost of manpower. On the other hand, limited manpower and an abundance of the basic raw material suggests that the minimum amount of work should be carried out on the material.

Material prices are largely governed by availability and the human effort necessary to provide them. Availability which seriously affects price stability is frequently overlooked by the designer. Material availability can be influenced, for example, by political situations, such as the shortage of copper in the United Kingdom resulting from the Rhodesian Crisis of the late 1960's, or by the seasonal fluctuations which occur with materials such as timber or natural rubber.

With materials found on the commodity market, their price reflects the degree of availability. In order to minimise the cost of the final product the material buyer has to be very skillful in purchasing. He must purchase when prices are low and in sufficient quantities to meet the company's requirements until the prices are next down; to understock would be to create shortages and to necessitate buying at inflated prices; to overstock could result in financial losses to the company due to the capital tied up in surplus supplies. A designer, before utilising such materials in large quantities, would do well to study their past stability and to take advice on their future prospects. Most good newspapers, having a financial section, publish the day-to-day fluctuations in commodity prices.

A designer may often create his own material shortages by using in quantity a material for which there is normally only a small demand, by increasing the demand a shortage leads to a consequent rise in price. Consider, for example, the design of thermal storage heaters which utilise off-peak electricity to heat blocks of refractory material for later emission. A typical heater core would require about 100 kg of refractory material and a manufacturer might produce about 3000 units per week. Clearly, before any new materials are considered, the designer must satisfy himself that they are available in quantity, to avoid supplies drying up at times of booming sales.

An integral part of material cost is the material which has to be 'thrown away', in other words, waste. The cost does not stop at the value of the material, but involves a waste-disposal problem (waste normally takes up more space than the parent material) and, rather paradoxically, the cost of manpower and machine time required to create the waste. The advantage of any moulding process is the little waste created, and any that there is, e.g. runners, risers and flashes, can be quickly recycled through the melt.

Waste can be minimised by studying the forms in which a material is available; only a fool would mass produce tubular components by machining out a circular bar, but many designers ask for tubes to be machined to specific dimensions for no particular reason.

Summarising:

(i) Stability of raw material prices and their continued availability are crucial to a successful product.
(ii) The cost of producing waste material can add substantially to the cost of a badly designed product.

3.9 SPECIAL TOOLING

The need for special tooling arises in one of two ways: either the product is being manufactured in large quantities and the use of tools expedites production, or the product cannot be made without them.

Special tooling for one-off products may be required for assembly purposes, where one component has to be located accurately relative to others whilst a joint or fastening is made. Two obvious examples are: the frames to support a brick arch and the hull of a boat during construction.

By skillful design the designer may be able to eliminate the need for the special tool altogether, but he must compare the economics of the elimination and use of the tools. Consider the frames required for the building of a wooden hull. If the frames were to become an integral part of the hull itself, they would no longe be classed as assembly fixtures. But in so doing it may be necessary to make the frames from a superior quality wood. They would probably be heavier than the frames required to maintain the shape once it had been formed, and hence add to the weight of the boat. These aspects have to be carefully assessed before the final decision is arrived at. One form of special tool required to make one-off products is the mould-maker's pattern; for this reason it is sometimes cheaper to avoid the use of castings when quantities are likely to be small. Plate 3.1 illustrates one such example, namely that of the large marine diesel engine. The traditional method of manufacture was to produce the cylinder block, crankcase and sump by assembling individually cast units. Since the demand for one standard marine diesel engine is relatively small, the manufacturer is compelled to stock a full range of patterns, the cost of which has to be borne by a few manufactured units over a relatively long period of time. The system illustrated in Plate 3.1 was developed by the Welding Research Association. The function of the cylinder block has been considered in its most basic form, i.e. it has to transmit the reaction generated by the gas pressure from the cylinder head to the crankshaft and secondly it has to support the cylinders and contain the cooling water around them. In this design the cylinders are formed by the liners—a common practice with diesel engines. The reaction from the head to the crankshaft is taken by a steel frame, constructed from rectangular hollow sections welded to end pieces; cladding of the framework with steel plate contains the cooling water. The advantages of the design are that it eliminates the use of castings and that different sized engines can be constructed by using different diameter cylinder liners and by varying the lengths of the frame members. It might be thought that the noise transmission would be considerably increased but research has shown that much of the engine noise originates from the flexing of the curved portion of the cast crankcase.

The more general use of special tools is in expediting the manufacture of products to be made in quantity. As already stated in section 3.7, there is usually a direct correlation between the use of tools for this purpose and the skill of labour using them. It is unlikely that skilled manpower can be found in sufficient quantity to carry out tedious and repetitive tasks and, as a consequence, the tools have to be designed in such a way that they may be used by the inexperienced operator.

To justify the use of tools for quantity production it is essential to show that the cost of each individual unit manufactured is lower when produced using the tools than without. Simplifying the problem, it can be assumed that for an item made using convential workshop equip-

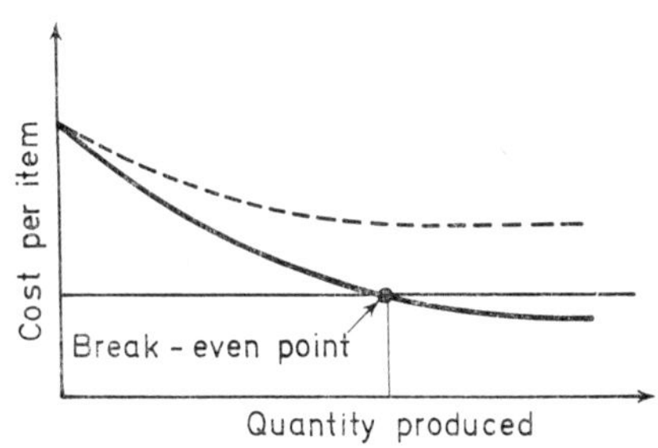

Figure 3.3 Cost per item againt quantity produced, showing the break-even point

ment the cost per unit will remain unchanged regardless of quantity. The assumption that the price per unit is constant may be represented as a straight horizontal line on the unit cost–quantity graph shown in Figure 3.3. If, on the other hand, special tools are used, then the cost of the tools must be apportioned between all the items being manufactured; in this way the cost of tools per item will be reduced as the quantity increases, giving the continuous curved line in Figure 3.3.

The point at which this curve and the straight horizontal line intersect is known as the break-even point. For a total quantity of items to the left of this point, special tools are not justified. The broken line shown in Figure 3.3 illustrates the case where special tooling can never be justified, even though the cost of tooling per item is reduced as the quantities are increased; the cost never falls below that using conventional workshop equipment. This may at first seem paradoxical, but it could be due to one of two situations; either the maintenance and replacement costs of the tools are high, or the tools rather than helping to speed up the work are adding to the complexities of the job.

How then can a designer take advantage of this situation? First, the designer must be aware of the quantities in which the product is likely to be required; this is an important parameter for the specification. He is then able to produce a design which makes maximum use of the production facilities available. Consider for example the case of glass-reinforced plastic moulding. This is a very attractive technique where complex shapes are required for small product quantities; here the glass mat and resin are laminated by hand, or layed up, the only special equipment being a simple mould. For large quantities it would probably be cheaper to use a steel pressing, but this has a number of disadvantages, e.g. being electrically and magnetically conducting, and prone to corrosion. If one of these properties is undesirable, the designer may revert to the glass-reinforced plastic moulding and the laying up by hand. If, however, he is fully aware of the quantities required and the production techniques available, he would probably adopt a hot-plastic pressing method; although the dies required may be expensive, the expense may be more than offset by the saving in manpower and the quantities required.

In a number of cases it is possible to take a finished item which is required in relatively small quantities and break it down into a series of identical standard units. Modular construction is a good example of this technique. The definition of a module is a standard unit having a fixed measurement. As a result, when modules are grouped to form an assembly, the dimensions of the assembly are exact multiples of those of the constituent modules. A simple example is the common house brick where, assuming that the bricks are not divided, the overall dimensions of the house will be multiples of the brick dimensions plus the layer of mortar.

The example shown in Plate 3.1 is a form of modular construction; here the module comprises an assembly of frame members and end fittings for each cylinder, which are then grouped to form the complete engine. Several identical corner fittings and frame members will be required and, since the corner fittings are cast steel, it is now worthwhile producing a pattern and making the castings.

Summarising:

(i) Special tools may be required to expedite the manufacture of a product required in quantity, or simply because it is impossible to make an item without them.

(ii) A break-even point must be established before special production tools are called for.

(iii) The designer should always seek to utilise standard units so that these units can be produced economically in quantity.

3.10 SPACE

Space is an expensive commodity, particularly when large unobstructed areas are required for assembly purposes. Space requirements can be regulated by timing and manufacturing techniques, but these in turn require certain provisions to be made at the design stage.

Some products are obviously space consuming, e.g. ships, aircraft, hovercraft, buses, etc., but some less obvious candidates begin to emerge such as furniture and food containers. Traditionally, a ship was built by laying the keel and building up the plating, bulkheads and ribs from this foundation. If the time from keel laying to launching was three years, then a large area of shipyard was tied up for the whole of this time; the only concession made to the economy of space was to fit out the ship after launching. The current trend in ship construction is to employ prefabrication techniques. In this way the large and valuable expanse of the slipway is only occupied for a relatively short time. Plate 3.2 shows a prefabricated section being lowered into position. The facility for prefabrication must be incorporated at the design stage as each unit must be structurally selfsupporting prior to its assembly.

The problem of the space taken by metal containers is again one of timing in the production cycle; the tin plate has a high stacking density whereas the finished product presents a space problem, occupying several hundred times the space taken by the raw material. This difficulty is overcome in some cases by making the fabrication part of the container-filling process; in this way the previously empty volume is now occupied by the packaged product.

Summarising:

(i) Large unobstructed assembly areas are expensive and should be occupied for the minimum time.

(ii) The timing of assembly can reduce space requirements but facilities to make this possible must be incorporated at the design stage.

3.11 SUPERVISION

The cost of supervision may be borne in one of two ways: either it is carried as an overhead as outlined earlier or, in situations where the supervision is clearly attributable to a product it may be costed directly, against that particular item. Supervision can be divided into three areas, namely: (i) general supervision of personnel, one of maintaining general discipline and orderliness, (ii) technical supervision of personnel; dealing with technical queries which may arise in manufacture, (iii) inspection of the finished product for defective workmanship and/or materials. Types (i) and (ii) are of a management nature.

Expense in supervision can result from both the quantity and quality of supervisory staff, which in turn may be directly dependent upon the skill of the operators. Design has little or no influence on the type of supervision defined in (i), except of course that an extremely large and complex project will require a large amount of manpower to see it through. On the other hand, the technical supervision of category (ii) is directly related to the complexity, or simplicity, of the design, particularly if the design deviates to any great extent from normal company practice. Any ambiguity in communication, i.e. incomplete drawings, etc., can lead to the need for interpretation, or misinterpretation, in the workshop. A major division exists between the two principal grades of inspection. When inspection becomes a matter for individual opinion, it is expensive; here the inspector has to be a person of experience, capable of reaching a carefully evaluated decision. If the inspection becomes a clear-cut accept/reject decision, based on a simple observation, then skill is no longer required and the only quality demanded of the inspector is one of integrity.

At the outset, the designer should consider the way in which the item is to be inspected. Take the case of a rod located in a hole. If the rod is to be a slide fit within the hole, the designer may indicate this fact by stating on the assembly drawing that a slide fit is required, or he may specify certain tolerances on the detail drawings which result in the desired degree of clearance. Where the parts are being made and fitted by the same man in a jobbing workshop, the former system of indication might prove preferable. Here there are likely to be no requirements for interchangeability of components, and several hours may be wasted attempting to manufacture the individual components to specific tolerances whereas the fit might be simply achieved by reaming, honing or lapping during assembly. Moreover, a jobbing workshop is more likely to be staffed by a skilled inspector capable of judging, by feel, the degree of fit. If, in contrast, the two items are being produced in quantity, some degree of interchangeability is desirable and, since the components may be made on an automatic or semi-automatic machine tool, it would be more economic to stipulate an acceptable tolerance on the drawing to be satisfied by the machining operation. Inspection can then be carried out by unskilled personnel using a simple GO–NOT GO gauge or a clearly marked comparator.

Summarising:

(i) Clear communication reduces supervision costs.
(ii) Inspection requiring a carefully evaluated opinion is expensive.

3.12 QUESTIONS

The following exercises are not aimed at producing a unique answer to a particular problem; the intention is to give the reader the opportunity to consider some of the everyday aspects of design economics. Where the book is being used as a text for a design course, these exercises may provide a guide for suitable mini-projects to illustrate this facet of design.

1. Find a piece of equipment suitable for export and determine its overall dimensions and weight. Calculate the overall density of the equipment and, by using the Appendix, decide upon the most suitable form of transport. By simple modification to these values, is it possible to make it more attractive to transport by other means? If the item selected is small, arrange it in crates of 10 or 100 units for the exercise.

2. Examine the interior trim of a motor car and make notes on the techniques employed to suggest expensive workmanship with a small financial outlay, e.g. the moulding of plastic to appear as stitching.

3. Select some components and decide whether any of the finishing processes might have been an integral part of manufacture; if not, could the item have been redesigned to this end?

4. High wages are often paid to unskilled employees to compensate for boredom. Find examples of unskilled personnel being employed on production work where this philosophy might not apply.

5. Compile a list of products which are space-consuming during and after assembly. Has provision been made in design to minimise the problem? If not, could any be made? What proportion of the final market price in each case could be attributed to this disadvantage? (Take account of the inconvenience to the trader and the stores department.)

6. Find examples of modular construction. Look at the definition of a module in a dictionary before doing so. How does modular construction facilitate manufacture in these cases?

7. Some one-off assemblies cannot be made without the use of special tools. Give examples and consider ways in which a modification might overcome the problem.

8. Estimate the cost of special tools required to manufacture a common domestic article. Compare this with the likely cost of materials and the labour in making it. Plot a graph similar to the one in Figure 3.3 and find the break-even point. Accuracy is not critical but correct orders of magnitude are.

9. Raw material is available in different forms, e.g. tube, round bar, rectangular bar. Select a material for which you have data and list the forms in which it is available. Against each write down the components which might be made from that particular form in order to minimise material waste.

10. Take any readily available engineered product and consider the ways in which it might be packed for export. Would a minor design modification simplify packing? Could the article be improved and the degree of protection required reduced?

11. Using common domestic products, such as washing machines or motor cars, find out roughly the weekly quantities that a single manufacturer is likely to turn out. If a penny, cent or centime, etc., can be saved on a single unit, estimate the total saving to the manufacturer over a full production run.

4 The Digital Computer as a Design Aid

4.0 ITS RELATIONS TO OTHER DESIGN AIDS

Traditionally the designer's 'tools' have been the drawing board and tee square, or more recently the draughting machine and an array of drawing instruments. On a broader scale, information is required from: design data sheets, associated drawings, standard parts lists, formulae and technical references; these are all essential in complementing a designer's experience. On the other hand much of a designer's work involves detailed and sometimes precise calculation. The favourite standby is the slide rule, but where precision is required the desk calculator has found a role. Many designers now have access to a digital computer, but this is often seen as a larger version of the desk calculator; in fact a computer may supplement many of the design aids already listed and lead to greater efficiency, but wrongly applied it may have the reverse effect.

The purpose of this chapter is to make the reader aware of the full capabilities of the computer in order that he may realise its true potential and some of the pitfalls associated with it.

Figure 4.1 shows a family tree of uses to which the computer may be put and the ways in which these are related. The three principal divisions are: calculator, data bank and draughting aid. By combining the latter two groupings, automated systems design is brought about. The following sections provide a more detailed explanation of the various branches in this tree. Although it may not be essential for the designer to fully understand what is going on within the machine, some aspects of computer construction and operation have been included in this chapter to give a more complete picture.

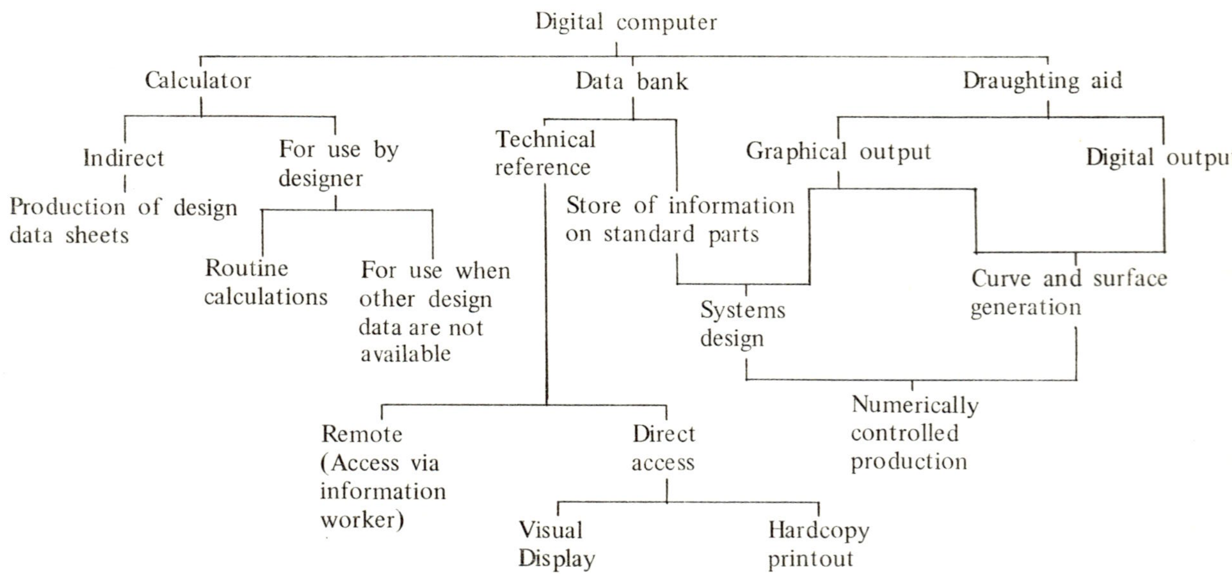

Figure 4.1 The relation between the possible uses of a digital computer as a design tool

4.1 INDIRECT USES AS A CALCULATOR

Much of the data used by the designer is to be found in the form of technical data sheets. The data are contained in codes of practice issued by the various national standards organisations, professional bodies, research associations and bodies specifically set up for the provision of design data such as the Engineering Sciences Data Unit[4]. Traditionally much of this information was based on practical experience and on experimental tests linked by empirical relationships; whilst this is one of the best ways of obtaining design data it is also very expensive and time consuming. Some of the early data sheets merely presented the results of formulae in a graphical or tabular form to eliminate the task of the designer having to calculate it for himself. The advent of the digital computer has made possible the analysis of relatively complex situations, containing several design variables, and has enabled these results to be presented in a graphical form appropriate to the designer (see for example reference 10).

As far as the designer is concerned he is handling just another set of design data sheets, and may be unaware of their origins. His confidence in using them stems from the confidence he has in the organisation which prepared and circulated the information. Because of this data sheet interface between the designer and computer, it is unlikely that the computer will intrude upon the normal design activity.

Summarising:

(i) The data sheet insulates the computer from the design activity.

4.2 USE AS A CALCULATOR BY THE DESIGNER

Situations often arise where calculations exceed the normal capabilities of the slide rule, either because they are likely to be complex and time consuming or because several alternative arrangements have to be computed and compared to arrive at the optimum. Here the designer must either make use of standard programs which have been prepared for him or he must write a program to perform the necessary calculations. In these situations it is very easy for the designer to lose sight of his principal task, i.e. that of design, and to find himself becoming involved in relatively complex and detailed calculations.

To put the computer into perspective as a design tool, it is necessary to see it in the light of the design process. Here it has basically two roles: the first is in the evaluation of the information appropriate to the problem specification, and the second is the calculating of values during synthesis. In addition it may also be used for data processing to justify the requirement, e.g. processing the results of market surveys, but this is generally outside the designer's control.

As already mentioned in Chapter 2, the values appropriate to the specification are such quantities as: load to be carried by a structure, power transmitted through a gearbox, reliability, and environmental considerations. If the design is a subdivision of a system, then the values may have been computed in the synthesis of the system. Often the quantities appropriate to the specification are simply arrived at, e.g. a static load applied to a structure might be simply defined but its distribution within the structure may be complex. On the other hand an acoustic load applied to a similar structure could itself require a considerable degree of analysis before the structure is even considered.

The computer cannot be used as a calculator at the speculation stage. Whether or not the digital computer might at some later stage provide a form of stimulus to the designer remains to be seen.

It is during evaluation that some designers, who have direct access to a digital computer, lose sight of what they are out to achieve. It will be recalled that evaluation commences at a superficial level and alternates between technical and economic considerations. At this level the type of calculation that can be done on the back of an envelope is sufficient to determine orders of magnitude for the relevant quantities. The next step in the evaluation requires a slightly more detailed look at the physical quantities, and this can be achieved by the use of approximate formulae and graphical interpolation. When the difference between the remaining alternatives is marginal the computer can be brought to bear upon the problem, to provide a detailed and accurate comparison, not necessarily carried out by the designer, but often by co-opted consultative bodies. Indeed the availability of a computer program to analyse one of the alternatives may rightly be instrumental in the final selection. Take the situation of a one-off heat exchanger required for a nuclear reactor. If a program that is capable of analysing a particular form of heat exchanger exists, then there may be every incentive in adopting that particular design, since quick and reliable predictions can be made about performance. If a similar heat exchanger is to be produced in quantity for use in power stations, it is essential to select the most economic design at a preliminary stage and then to consider methods of assessing performance. Unfortunately, the designer sometimes becomes obsessed with a program's availability and as a consequence fails to recognise the merits of a more appropriate design solution.

Synthesis is the stage at which the digital computer finds greatest use in design. The analysis should first be explored using standard design data, such as the graphical or tabular presentations of information, and should only be transferred to the computer for one of the following reasons:

(i) Design data are not presented in any hardcopy form.
(ii) Accuracy cannot more easily be achieved by other means.
(iii) A computer's speed of operation makes it possible to carry out lengthy calculations in a relatively short time.

The reasons listed in (i) to (iii) are best illustrated by examples.

Many complex computer programs have been written which accurately describe the performance of a particular component or subsystem. Because of their complexity they are expensive and time consuming to run. Although it would be possible to calculate the performance of the item under several different conditions it would be extremely expensive; since the cost of running the program has to be borne, the price must be passed on through the sale of the resulting data sheet. Such an exercise would price the data out of the information market. One policy is for the owners of the program to offer a user service. The designer provides the service organisation with the appropriate operating data for the particular configurations and by running the program for this specific set of data the charge is only for the single run. The difficulty with this arrangement is that the designer must first find a consultative organisation possessing a program capable of solving his problem. A number of computer bureaux offer such services, and in the United Kingdom the National Computer Centre at Manchester holds records of available programs but, in spite of this, searching for the proper source can be both time consuming and sometimes fruitless. An example of programs in this category are those for the dynamic performance of fluid-film bearings. Whilst the designer is able to predict the characteristics of a bearing under conditions of steady load to see whether it fulfils his need, to carry out a dynamic analysis the bearing cannot be studied in isolation from its surrounding structure and rotor. To provide design data sheets catering for all the possible structural configurations would be a formidable task and as a consequence it is easier to compute a specific configuration as required.

Reason (ii) still requires initial approximate calculations to be made before a full analysis can be justified. Although the digital computer is capable of performing very accurate calculations, it is only as accurate as the data it is handling and the program it follows. A program is a mathematical representation of physical conditions and as such no program can exactly describe a physical situation; some degree of error must be accounted for. A consideration often neglected by the designer is that the percentage error may change with the numerical magnitudes employed. Much of the documentation provided with standard computer programs fails to make any mention of accuracy. When such references are made they often tend to be somewhat optimistic, which is natural when it is realised that the accuracy reflects, in part, the abilities of the person who has developed it. All numerical representations of physical situations have a limited range over which they are valid. This may be due to a gradual divergance between the mathematical model and the actual event or a change of physical state which is no longer represented by the numerical analysis. A good example is the analysis of gas flow through a small slot. When the size of the slot approaches the molecular mean free path of the gas, the gas can no longer be treated as a continuum and as a consequence methods of analysis which treat the gas as a Newtonian fluid can no longer be applied. This very problem arises in the design of fluid film bearings employing helium as the working fluid. Any computer program representing a physical state and for which limits of validity are not stated must be considered suspect.

Reason (iii) is probably the principal reason for the digital computer's acceptance in engineering design. The emphasis should be on the term 'lengthy calculation'; the reader should not take this description to imply complicated or sophisticated, indeed the individual evaluations it carries out are of a very elementary nature. However, many complex analyses can be broken down into a very large number of trivial calculations. Normally it is not worth man's while tackling the problem in this way, because of the vast amount of time required for such calculations; it is therefore more efficient for him to use his intellect to arrive at a more sophisticated solution quickly.

Consider a formula containing one independent variable and one dependent variable. If differential calculus were not available, the only way of finding the maximum value for the dependent variable would be to evaluate the formula for several values of the independent variable until the maximum is found. Obviously the modern student would differentiate the expression and set the derivative equal to zero to obtain an expression for the maximum. It will be seen later, in the sections on optimisation, that it is not always possible to seek the aid of differential calculus for this purpose and a more sophisticated form of the former method has to be adopted; here the computer can be used to carry out a multitude of calculations to find the maximum.

Another type of problem is that where an expression has been derived to describe a certain design situation, it is then found that this expression cannot be solved by normal analytical means, and a system of 'trial and error' has to be adopted so that as the trials proceed the errors diminish to zero when the actual solution has been found. This is known as an iterative solution and is illustrated in Example 2.

Summarising:

(i) The computer is normally only really effective in the specification and synthesis stages of design. Its use for evaluating alternatives could lead to unnecessarily detailed calculations in the early rounds of assessment.

(ii) Instructions for the use of existing programs should be studied to examine their limitations. If no reference is made to limitations the program should be considered suspect.

(iii) Computers can only be used for problems which can be broken down into a series of trivial repetitive calculations.

4.3 THE COMPUTER AS A TECHNICAL REFERENCE BANK

As well as being a calculator the computer can have a large capacity for information storage, and by human standards the information is retrievable in a very short time.

There are basically two ways in which the computer may store information. In the first, the information is stored within the core of the computer itself, in what is generally a ferromagnetic core; such stores are, however, very expensive and are normally reserved for the storage of numerical data required in calculations, and for the program software. A less expensive storage medium is magnetic tape which is in effect a very sophisticated form of tape recorder linked to the main computing unit. The tape stores a series of impulses as opposed to speech or music. Although magnetic tape is relatively cheap, it carries the penalty of being slow compared with a magnetic core. Another form of magnetic store, intermediate between core and tape, is the disc. The magnetic disc roughly resembles a long playing record but the surface contains no groove and it is coated with a magnetic film similar to that of a tape. The readoff heads are placed side by side along a radius and very close to the disc surface. The disc normally runs continuously, meaning that one of the heads will 'see' every position on the disc surface once per revolution; the higher the speed of revolution the faster the retrieval of information. Several discs are normally placed on the same spin axis, with spaces between for readoff heads.

The type of computer used primarily for the handling of information is known as a data-processing computer and usually comprises a relatively small core store, for holding the control program, together with large amounts of tape or disc storage from which the data is retrieved. It is difficult to generalise on computer facilities and the reader may well find a time-sharing computer with a large core capacity being devoted to information retrieval, particularly if the machine is also being used for some other design function.

It now remains to be shown how the digital computer can be used as a data bank. As a simple store for designer's data the computer is very inefficient; although the data may be stored in a very compact form it is not as acceptable to the designer as a well indexed comprehensive reference book. The principal advantage of the computer in this role is its ability to search for documents and references about which it has information, and which may not be readily available from a common source. Having knowledge of the particular reference the designer is then able to use one of the other media such as a library containing the original document, microfilm, or even a graphic display linked to a computer to view the information contained in the reference.

Most information retrieval methods are based on a system of keywords, which are grouped together to form what is known as a profile. Each document which is to be referenced by the computer data bank has a descriptor ascribed to it; this descriptor comprises a string of keywords which describe the contents of the document. A somewhat superficial though common method is to select keywords or equivalent keywords from the document title. A more rigorous method is to use the abstract, or summary, of the document to find suitable keywords to describe the contents. The ultimate, of course, is to have the indexer read the full report before selecting suitable keywords. Whilst this would be the most desirable method, time rarely permits such a detailed study. These keywords describing the work are stored within the computer together with the document title, date and source, etc.

On the other end of the system is the designer requiring the information. The first task in retrieval is to compile a profile of keywords which adequately describe the information required. These keywords are then fed to the computer which matches the keywords against the keyword descriptions for all the documents it has referenced.

To provide a uniformity of definition, all the allowable keywords are listed in a thesaurus. The topic for which the information is required must be described by a profile containing only keywords taken from that particular thesaurus. In order to narrow down the field of search the matching is only carried out with groups of keywords. This point is best illustrated by way of an example. Consider a designer requiring information about the depositing of carbon for the manufacture of a carbon resistance potentiometer. A suitable group of keywords for describing this topic might be: POTENTIOMETER+CARBON+COATING. This group, however, would only provide references containing information about the carbon coating of potentiometers It may be advantageous to exclude POTENTIOMETER from the group as any method of carbon coating would be of interest to him, leaving: CARBON+COATING. On the other hand, this may provide information on carbon deposits in internal combustion engines, which is of no interest at all. In some systems the logic can be extended to exclude certain terms in which case the following group might result

CARBON+COATING excluding COMBUSTION. Several groups can be arranged in this way to provide a single profile.

The most common arrangement for obtaining the information is for the designer to use the system in a remote sense. Here the profile is compiled and then deposited with an organisation where the search is to be carried out. The alternative arrangement is for the designer to have direct access to the computer via a teletype terminal or some similar device. The advantage with the former arrangement is that the designer can be advised on the construction of his profile by the information officer carrying out the search. The drawback is that due to the time taken for the search to be made and the results returned; the profile must be very accurately compiled the first time, otherwise valuable time is lost in modifying the profile and waiting for the result. Whilst direct access is expensive to implement, and it also deprives the user of the professional advice, it does enable him to experiment with his profile and get a 'feel' for the system.

This is particularly important in information retrieval, as a profile in one subject area may yield only a few references, whilst a similar profile for a different topic may well swamp the engineer with a deluge of unwanted information.

When used in a remote way the designer will be provided with a hardcopy (paper) form of output, but for direct access the output can be in the form of a type written paper copy, or an alpha-numeric display on a cathode-ray tube (CRT). The advantage of the CRT display is the 'instant' way in which the information can be presented. With many systems the user can continuously modify his profile, observing the result on the CRT display, and when satisfactory he can give an instruction for the high-speed printer, at the computer installation, to print out the details which will be sent through the normal postal services.

Summarising:

(i) The computer should not be looked upon as an alternative to the book for storing the written word. Its strength is in its ability to rapidly sort short groups of words or characters to provide details of technical references for the designer.

(ii) For retrieval based on keywords, the keywords must be carefully selected, grouped and then tested in order to avoid vast numbers of unsuitable references.

4.4 USE AS A DRAUGHTING AID

A very necessary part of design is to communicate ideas to a manufacturer or to the designer of the subsystem. Many companies have reduced the amount of time wasted in orthodox drawing by introducing freehand sketching, symbolic representation, stick-on symbols and standard drawings of parts only to be dimensioned and materials specified. Improvements of this sort could not easily be implemented in lofting (the laying out of curves and surfaces). The problem was most apparent in the aircraft and shipbuilding industries where the process was both costly and time consuming. Since the basic curves of a fuselage, or hull, have to be specified before work can be started on designing the structure, it was necessary to add at least six months to the lead time required on a project of this sort. To lay out the most basic of lines for a medium-size airliner it could take as many as thirty loftsmen to complete the work. The process of curve fitting is essentially one of geometrical interpolation and the work is ideally suited to the interpolation techniques available on a digital computer; more will be said about curve fitting in later sections. This led to the exploitation of the computer's capacity to deal with graphical manipulations in other fields of engineering design.

To understand the computer's potential it is necessary to examine the ways in which the designer can feed information to the computer and the devices available for providing the output. Broadly, the input and output media fall into two general categories: digital and graphical.

4.5 DIGITAL INPUT AND OUTPUT

The traditional, and still the most common, method of input and output is the digital form. The computer is provided with a series of decimal numbers and after executing its program produces the results in the same form. Methods for conveying this information to the machine fall into two general types: holes punched in either a card or a paper tape or the use of magnetic tape to store and convey the digits, and systems where the user feeds each digit directly to the computer. The former method is known as batch mode, i.e. the digital information is accumulated on paper tape, card or magnetic tape, and is supplied to the computer as a batch of information, whereas the latter is known as on-line mode since the machine receives each digit of the data directly from the user as it becomes available. These two alternatives are shown diagramatically in Figure 4.2.

With either of these methods the designer must have access to some device capable of translating the digits. The simplest and most common form is the card or paper tape punch, comprising a normal typewriter keyboard which actuates the punching mechanism to produce the holes in the tape or card. A similar device instead of punching the tape leaves an impulse on a magnetic tape. An on-line terminal is in effect a typewriter keyboard connected directly to the core of the

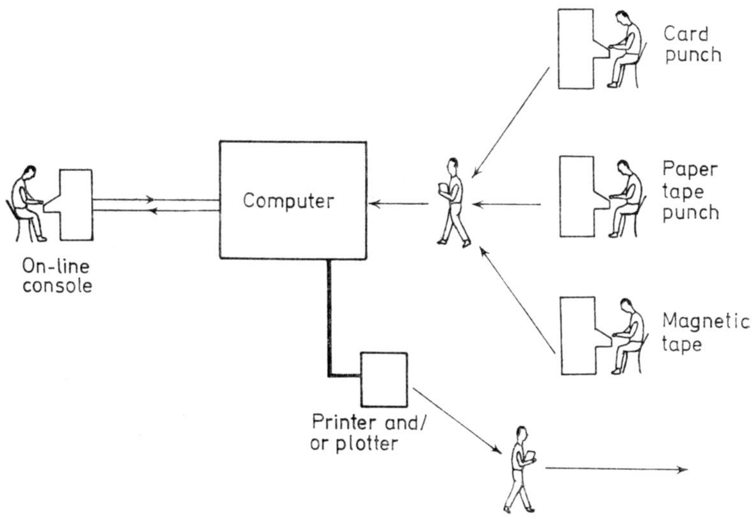

Figure 4.2 On-line use of the computer (left) contrasted with batch mode (right)

computer. Whilst all geometrical information can be presented in a digital form, it is not always convenient to measure coordinates from an existing drawing and then convert them into numerical data. To overcome this problem the digitiser was developed; a typical digitiser is shown in Plate 4.1. The array of equipment is necessary to generate the tape which is supplied to the computer. What appears to be the draughting head provides the coordinate reference position for the digitised information. The equipment is used as follows: the drawing to be digitised is placed on the table. On the head is a pointer which is moved to the appropriate position on the drawing. When the pointer is located, a button on the keyboard is pressed which passes the coordinates locating that point to the magnetic or paper tape. Digitisers are extensively used for producing numerical control tapes from existing scale drawings.

Methods of input have been briefly covered, but what about output? The normal method of outputting data from the computer is by line-printer. This is a high-speed printing device which in effect prints a complete line of type at a time. For draughting, a list of numerical coordinate positions is of little or no use. The most acceptable form of design output is that produced by a numerically controlled automatic draughting machine, often known as a flat-bed plotter. These machines are relatively slow in operation and one of the 'bottlenecks' in high-speed computing is the output device. To overcome the problem, the computer is made to produce a magnetic tape containing the coordinate information; this is then read by the draughting machine at a more leisurely rate. Plate 4.2 shows a typical installation. It is important to note the size; although smaller machines are manufactured, large tables are essential for generating full-size layouts of aircraft, ships and cars. A carriage spans the width of the table; this is driven along the table to provide one of the ordinates, whilst a smaller carriage containing the pen traverses the width to produce the other ordinate.

The digital information provided by the computer only defines discrete points. The simpler draughting machine merely connects these points by a series of straight lines, but more sophisticated units have facilities for fitting a smooth curve.

Sometimes the designer wants a much smaller drawing, similar to that produced by manual draughting means, often containing dimensions and notes for manufacture. Such drawings are more efficiently produced by what is known as an incremental plotter. These plotters generate lines on the plotting medium as a series of discrete steps; one line of the step being parallel to one ordinate and the other line parallel to the second ordinate. Although these steps are small and precise they are clearly discernible to the naked eye. Some plotters have a facility which enables increments to be made in directions at 45° to the two plotter axes, having the effect of a slightly smoother curve. Plate 4.3 shows a typical incremental plotter. The machine produces a continuous strip output of the plotting medium, in this case paper. This is perforated with sprocket holes along its two edges. Movement along one axis is produced by driving the paper in and out of the

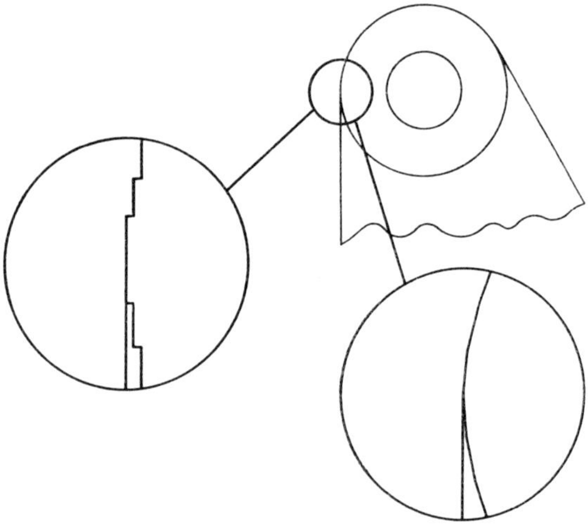

Figure 4.3 An enlargement of, on the left, the stepped line typical of an incremental plotter and, on the right, a line obtained from most numerically controlled draughting machines

machine. Above the paper leaving the machine can be seen the pen carriage. Movement of the carriage provides the second axis. In Figure 4.3 the output from the incremental plotter is compared with that of the draughting machine. Note that the incremental plotter only produces a truly smooth line when the line is parallel to one of the axes.

The incremental plotter is fast and versatile. Because of their higher speed they are sometimes connected on line to the computer. Since the paper is produced from a continuous roll, there is no need to reload for every drawing. A programmed instruction merely lifts the pen whilst the paper is advanced to the next drawing position. Furthermore, the annotation facilities provided as standard with plotter software enable labelling and dimensioning to be carried out in an economical manner.

Summarising:

(i) Digital input and output is of little direct use as a draughting aid. It is more general to use a digitiser to input numerical information about shapes, and to use the digital ontput to drive some form of automatic draughting machine.

4.6 GRAPHICAL INPUT AND OUTPUT DEVICES

Because digital input and output devices are slow, they are rarely connected directly to the computer and are more generally used in a batch mode. The alternative is the use of a graphic display. What follows, although not essential to the use of visual displays, should help the designer in his understanding of the equipment.

Graphic displays are normally based on the cathode-ray tube (CRT), giving a line diagram on what appears to be the screen of a television. The display differs from that of the normal television in that a television picture is produced by sweeping the entire screen with an electron beam. By modulating the intensity of the beam during this sweep it is possible to produce a picture made up of horizontal parallel lines. This is known as a 'Raster' scan. Since the graphic display from a computer is only concerned with discrete lines, it is no longer necessary to continuously sweep the whole screen; the beam merely traces out the lines on the screen. It traces out one line and then moves to the next and so on. A permanent record can be made of the picture by photographic means.

Having established a pictorial display, some means of modifying it must be found. The most common system combines the manipulation of a tracking cross with a light pen. The cross comprises two short lines, one vertical and one horizontal. It can be moved to any position on the screen by the tip of the light pen and provides a pointer for indicating the part of the display which is to be modified. Plate 4.4 shows a typical graphic display console. By using the control keyboard the operator is able to instruct the computer to find a specific set of data and to display the data on the screen in a particular form. He may also use the keyboard to remove and store the picture, or to output it in a different form such as paper or magnetic tape, a line-printer listing of coordinates, or a paper copy produced on an automatic draughting machine. The function keys represent a particular task which the designer may wish to perform, e.g. delete, draw a straight line, draw an arc, etc. They are often described in a group as the 'button box'.

Before considering the method of operation it is well to give some simple examples of how it might be used. Suppose a picture is already displayed on the screen and has to be modified in some way. The designer wishes to delete a line; by pointing to the line and pressing the function key assigned to delete, the line in question disappears. He may then wish to place a radius at the intersection of two lines; by pointing to the lines and pressing the function key representing an arc tangential to the two, the radius will appear, but what about the magnitude of the radius? The way in which this is defined will depend on the console system being used, and on the software controlling the console. In some systems the radius might be given by punching in a numerical value from the keyboard or, for radii such as those used in bendding sheet metal, one of the function keys might be assigned to the standard radius. An alternative arrangement is to display a string of digits along the edge of the screen; the designer then points to the digits in turn with the light pen and by pressing the interrupt key on the button box the appropriate radius would appear.

The function keys on the button box do not have specific meanings in themselves. The function performed by pressing the key is assigned to that key by the control program and an individual key's function may be changed from program to program. To avoid confusing the operator it is desirable to maintain a degree of consistency within a particular system.

Whilst the use of button box and light pen is the more common form of display manipulation, other methods exist. One such system involves the use of the light pen and what is known as a 'menu'. It has already been explained how digits may be displayed along the edge of the screen, but it is also possible to display functions on the screen in place of function keys. The functions may be displayed in a descriptive or pictorial form; the function keys are then replaced by a single interrupt key. A user first points to the area of the display which is to be modified, and then to the particular function on the menu after which he presses the interrupt key to perform the modification. The advantage of the menu and light pen method is that it is psychologically more acceptable to the new user since all the manipulation is carried out on the screen. The major disadvantage is the substantial area of screen permanently occupied by the menu and in a sense wasted. Furthermore, an experienced operator can work more quickly with a button box and light pen.

Other less common systems involve a cursor in place of the tracking cross, which is controlled by either a 'joystick' rolling ball, or 'mouse'.

The path of the CRT electron beam is given by a digital sequence of X and Y ordinates from the computer. The action of the electron beam striking the rear of the screen causes the phosphor coating at that point to glow. This glow has a short but finite duration, and if the

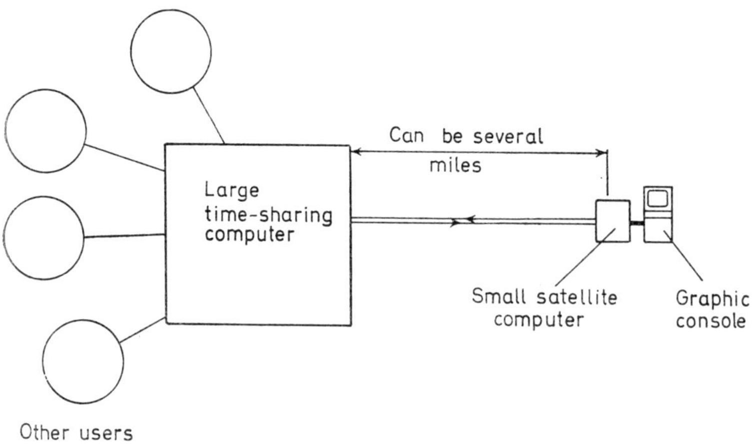

Figure 4.4 The diagrammatic layout of the graphic display console, its picture support computer and the larger time-shaving machine

viewer is to see a continuous picture the beam must return to points on the line to reactivate the phosphors before the human eye can detect any decay in the glow, i.e. in this period the beam scans every line on the screen and returns to its starting point. If this is carried out at a frequency below 50 Hz the picture will appear to flicker to the annoyance, and ultimate fatigue, of the user. The computer, in order to maintain the picture unchanged on the screen, has to recalculate the position of every coordinate fifty times in one second; if the picture has to be changed, additional calculations must be carried out to modify the coordinate positions. Most present-day systems employ a small satellite computer to maintain the unchanged image on the screen, which in turn is connected to a large multi-access time-sharing computer which carries out the calculations necessary to effect any change of image. This is shown diagramatically in Figure 4.4. In the early days of graphic displays the console was normally linked directly to a medium-size computer, giving a number of serious drawbacks: the console had to be located in close proximity to the computer, otherwise the slow transmission rates resulting from the electrical capacitance of the very long cables caused the picture to flicker; as the same computer was used to support and modify the picture, any major modifications could result in the loss of a picture for several seconds; much of the computer's capacity was lying idle whilst the user was thinking, the alternative being to site the console at the centre of a computing complex, rather than in a technical or design office, so that the spare capacity could be used for other purposes.

The light pen appears to be more a piece of magic than an engineering development. The principle of operation is, however, very simple to understand. The pen contains a light sensor at its tip with the console lead connected to the opposite end. The glow from the phosphors, representing the tracking cross, are detected by the light pen. The pen has a finite field of detection, represented by the circle in Figure 4.5(a), and it is possible for the computer to detect whether the cross is centrally placed within this field. If the pen is being moved,

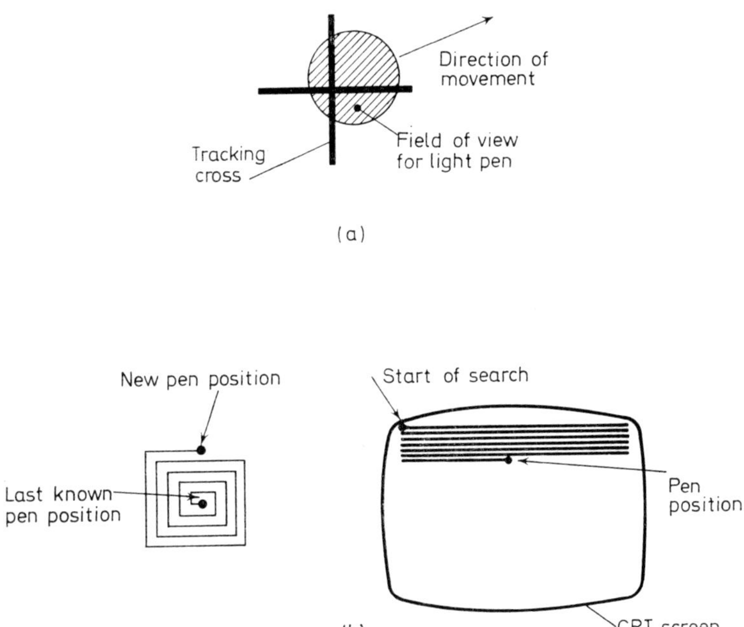

Figure 4.5 (a) Enlarged view of tracking cross; (b) The two ways of finding a lost pen position—on the left the expanding spiral method and on the right the application of the raster scan

the cross can never be in this central position, therefore the computer is constantly checking and trying to centralise the cross and light pen; as a consequence the movement of the cross normally follows the movement of the pen. If, however, the pen loses the cross, due to rapid movement by the user or by the pen being removed from the screen, it is then necessary for the computer to search for the pen's new location. This can be done in one of two ways shown in Figure 4.5(b): either by starting at the last known point and driving the beam in an increasing spiral until the new position is detected, or by carrying out a raster search whereby the beam scans the whole of the screen along horizontal lines until the pen is found.

An excellent, detailed explanation can be found in reference 11.

Summarising:

(i) As the picture must be renewed at a rate better than 50 Hz, the console must be located close to the picture support computer, which is normally a satellite to a central computer used to calculate modifications.

(ii) Draughting functions may be represented by the keys of a button box, or by a menu. The former method is usually faster in use but a menu may be psychologically more acceptable to a new user.

4.7 DRAUGHTING APPLICATION

There are basically three fields in which the digital computer can be usefully used as a draughting aid; these may to some extent overlap.

The earliest application was that of defining curves and surfaces, which will be explained further in the next chapter. The others are systems design and designs leading to numerically controlled production processes.

Frequently, the designer is faced with the designing of a system where many of the components are standard parts. To apply normal drawing techniques he either wastes much effort in the repetitive drawing of details or represents them by symbols. When components are represented by symbols two difficulties arise. The symbol must be easily recognised by the manufacturer, otherwise misunderstandings may result. Secondly the symbolic representation can in itself become complex and time consuming, an example being electronic circuitry; this contrasts with the simple symbolism of representing the position of rivets in a structure by centre lines. By using a graphic display unit and a data store of standard components, or functions, the designer can contsruct a system comprising a number of standard components relatively quickly.

An obvious condidate for this design automation is electronic circuitry and it was here that much of the pioneering work was carried out. The symbolism of an electronic circuit is in itself simple, but the drawing of each individual component in the circuit is time consuming. By using the graphic display, the designer is able to build up a circuit by pointing to a certain position on the CRT with the light pen and then calling for the image of a standard component to appear. Information about the component is stored within the computer. This information comprises the graphic details required to produce the line image, the component number and information about performance. As the complete circuit comprises standard components it is possible in many cases, from individual component performances, to have the computer describe the overall performance of the circuit. In this way the designer can make rapid modifications to the circuit and test the predicted performance before deciding whether or not further modifications are necessary.

Having established the design on the CRT screen, the full information about the circuit can be stored within the computer and is retrievable in a number of forms. By using an incremental plotter a hard-copy circuit diagram can be provided for use by an engineer constructing a circuit in the traditional way, or for general reference purposes. With the advent of microminiaturisation many circuits are produced in a printed form. Here the computer can generate numerical tapes from the stored data for the control of optical mask-cutting machines required in the manufacture. In addition, the lineprinter may produce schedules of components required in the circuit.

A similar process in the field of mechanical engineering has been adopted by British Leyland Limited for the design of blanking tools. The profile of the sheet to be cut by the tool is established and is displayed on the CRT screen.

To produce the die from a single block of tool steel, by machining out the profile on the female unit, would be wasteful of material and would create problems in the precision machining of the inner faces. To overcome this difficulty the die is made up of individual tool-steel segments clustered together to form the full profile, the whole being located on a base plate. Each segment is machined to produce that particular section of the profile. The segments must, of course, be precisely located relative to each other to avoid discontinuities of the profile at their boundaries. Each segment starts out as a rectangular or triangular block. These are laid side by side in such a way that the profile never falls outside one of the blocks. In this way it can be seen that the section of profile to be machined into each block has been defined. Having established a suitable distribution for these blocks, other items can be fitted into the arrangement such as dowels for location, clamping devices for the blocks, guide bars for aligning the dies, and locations for the stripper plates. Since these would normally be standard components, some rearrangements of the tool-steel blocks may be necessary; the final design has to be decided on a trial and error basis.

To produce the dies by traditional draughting methods is time consuming and tedious. Using a graphic display console the designer tests the possible arrangements quite rapidly and by a process of trial and error arrives at an optimum arrangement.

He starts by displaying the profile to be cut on the screen. Information about all the standard components is stored within the computer. The components can then be called by code numbers and positioned on the screen by means of the light pen. Unsatisfactory components are quickly deleted using the light pen and function key. In this way a complete blanking tool is designed. The sequence is shown in Figure 4.6.

From the data stored within the computer, hardcopy drawings are provided to guide the toolmaker during construction; information is available for producing numerical control tapes to machine the profiles on the individual die blocks, and for the purposes of efficient storekeeping a schedule of components can be printed for each tool.

As an example of numerically controlled (N/C) manufacture, consider fabricated steel structures. A particular case is factory buildings, where the same basic configuration is employed over a range of differently sized units. A factory having a span roof can be made higher by increasing the lengths of the supporting stanchions, or wider by increasing the span; the configuration may remain unchanged with variations in span, but the sizes of each member would vary to meet the changed load requirements. Most basic structures can be simply calculated using established methods. A designer having an interactive link with a computer, is able to test many standard configurations in varying circumstances. The computer stores information about the standard members available to the company and standard methods of joining these members. By comparing alternative designs, in rapid succession, the designer is able to find the most economic way of satisfying the requirements. The standard structural arrangements and the jointing methods can be specified numerically.

For production, printed schedules can be provided for the cutting lengths of the members, and also the drilling arrangements necessary for various joining methods. This data can of course be made available in a tape form for the control of machine tools. Plate 4.5 shows one such machine used for drilling the ends of structural members to match the fastening arrangement. The positioning of holes in the two mutually perpendicular planes is controlled by the information on the tapes; when used in conjunction with an N/C cutter the only manual task is the painting on of the part number.

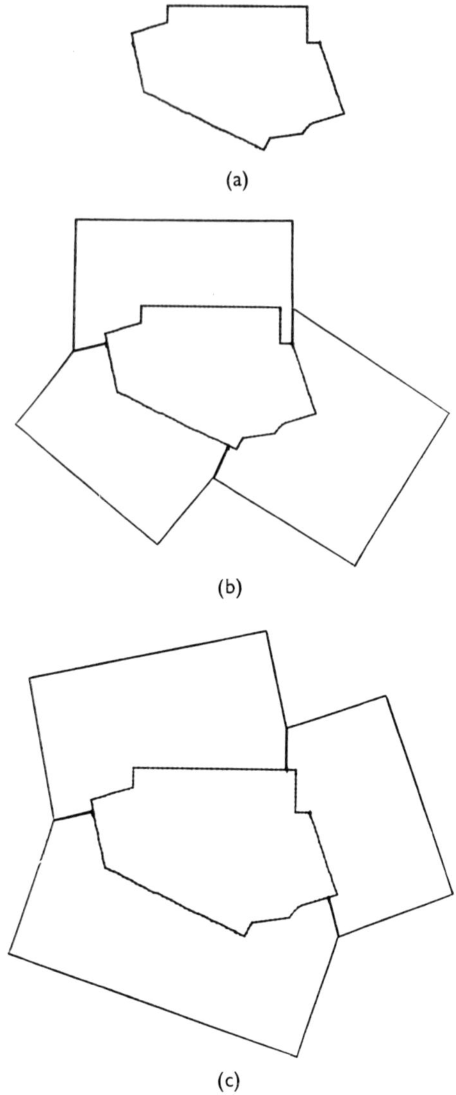

Figure 4.6 (Originally displayed as white lines on a black background.) (a) The basic form to be cut. (b), (c) Possible arrangements of the blocks around the form. The lower die of a similar tool shown in (d) gives the positions of dowels and clamping bolts for the blocks as an assembly drawing of the lower half of the tool. The upper die is illustrated in (e) and a section of the whole assembly in (f)

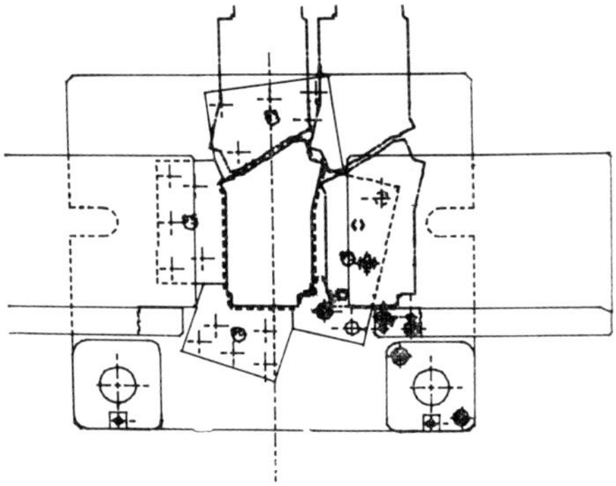

(d)

29 CONTINUE 26 CHECK PLOT
30 ERASE 27 SCALE
31 QUIT 28 SKIP

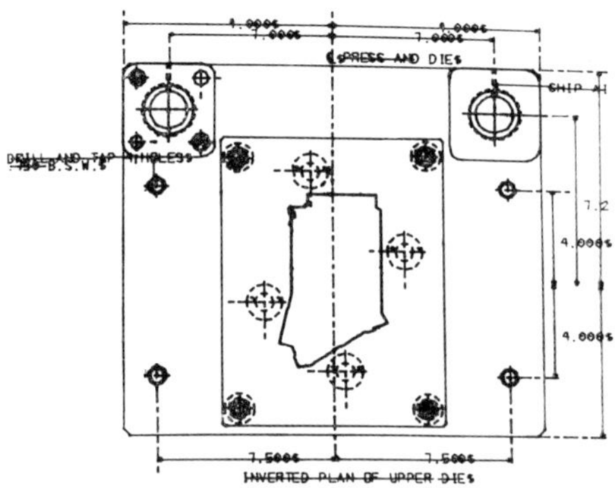

(e)

29 CONTINUE 26 CHECK PLOT
30 ERASE 27 SCALE
31 QUIT 28 SKIP

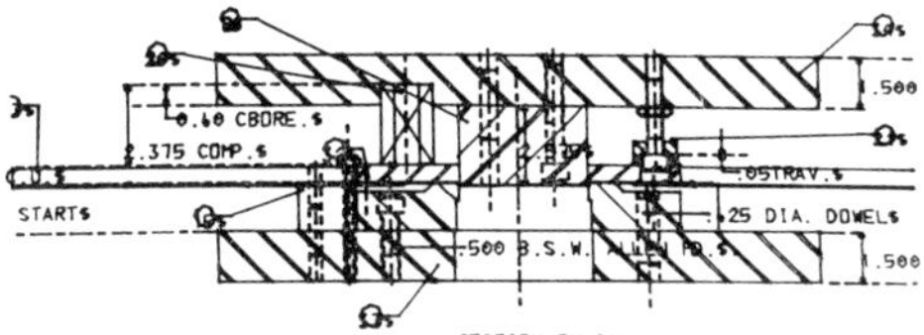

(f)

Summarising:

(i) The three principal applications as a draughting aid are: curve fitting, systems design and design leading to N/C production.

4.8 EXAMPLES

The following examples have been selected to illustrate the nature of computer calculations.

4.8.1 Shear Forces and Bending Moments

Consider the shear force and bending moment diagrams for a simple beam. Once the reactions at the supports have been found, the shear force diagram is determined by starting at say the right-hand end and moving an increment at a time to the left. At each increment the algebraic sum of the forces to the right of that point would provide the value of the shear forces acting at that position. The stupid student would follow this step-by-step process blindly, whereas the more alert individual would realise that there is a linear increase in shear force to start with and that there is no change in the value between the points loads. The computer being a stupid thinker could be programmed to follow through the monotonous step-by-step process, providing point values of the shear force at very small intervals along the beam.

In practice, points loads cannot exist; individual loads must be spread over a small finite area. Figure 4.7 shows diagramatically a sec-

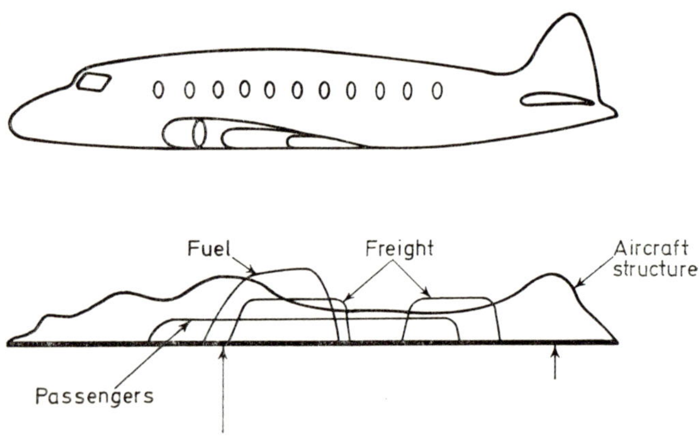

Figure 4.7 The type of load distribution which might be expected from a civil aircraft, represented as a complex distributed load on a beam. The supports represent the lift on the wings and the tail

tion through an airliner and the equivalent beam representation when it is in level flight. The distribution of the load is made up of passengers, freight, cabin equipment, fuel, navigational equipment ... and the aircraft structure. The approximate weight distribution for the structure can be determined by experience from earlier designs. The outcome is a nonuniform distribution of load along the beam, supported by reactions at the wings and tail. By considering a particular point on the beam and summing every element of the load to one side of this point it is possible to calculate the shear force at that position. Although the calculations are trivial, to repeat the exercise for the full length of the beam is a tedious operation which is ideally suited to the computer. Having established the shear force, a similar exercise can be repeated for the bending moment.

The level flight condition produces one form of shear force distribution but, for the complete picture, all flight and landing conditions have to be considered and their shear force diagrams superimposed to find the most severe shears on the fuselage. Once a more definite structure has been determined the whole operation can be repeated to take account of an improved structural load distribution.

This is an example of how the computer, by carrying out an enormous number of trivial calculations, can be used to solve a relatively complex problem.

4.8.2 Finding the Neutral Axis in a Complex Beam

For a simple beam having a uniform cross section along its entire length and made of a single material, it is a straightforward process to calculate, from the geometry, the position of the neutral axis. If, however, the beam comprises several different materials and has a complex cross section, which may vary along its length, an iterative (trial and error) method is useful, but this results in a multitude of repeated calculations.

The classical equation for the stress in a beam is

$$\frac{M}{I} = \frac{f}{y}$$

where M is the bending moment at that position in the beam (probably arrived at by a method similar to that in the previous example), I is the second moment of area of the beam about the neutral axis, and f is the stress at a position y from the neutral axis. By definition, the neutral axis is the position in a beam where the longitudinal stresses are zero.

The iterative approach is to assume a neutral axis, possibly through the centre of the section, and then to calculate a stress distribution in the section. The assumed neutral axis if correct should coincide with the position of zero stress. If not, the position of zero stress is taken to be a better assumed position than before. The stress distribution is then recalculated on this basis and, if the iteration converges, the new position of zero stress should this time be closer to the improved assumption for the neutral axis. The method is repeated until there is little or no change between successive positions.

Referring to the previous example, the beam section might be that of an aircraft fuselage. To calculate the second moment of area about the true neutral axis would in itself be time consuming, but to have to do it for every assumed neutral axis until the true one is found requires the speed and capacity of a digital computer.

4.8.3 Transient Heat Transmission

If heat is applied to one surface of a large sheet of material, some time must elapse before the temperature of the face on the other side has built up to a steady value. This is because the 'cold' material, as well as transmitting the heat, stores some. This phenomenon can be analysed by breaking the sheet up into theoretical laminations, each lamination being very thin. The analysis proceeds by constructing a heat balance for each lamination, i.e.

$$\text{Heat entering} = \text{Heat leaving} + \text{Heat stored}$$

By taking a small interval of time it is possible to calculate, from the thermal conductivity, the rate of heat entering a particular lamination of unit area and, in that time, how much will be stored and how much will be passed to the next lamination. The calculation proceeds lamination by lamination through the sheet until the temperature of each lamination for that instant in time has been calculated. The time is advanced by a further interval and the whole process is repeated. In this way the temperature growth with time can be observed throughout the material. Fuller details of this method can be found in reference 12.

The computer was used in this way to analyse the performance of domestic storage heaters, operating on off-peak electricity at night. This work is reported by Iqbal[13] and illustrates another example of a complex problem being divided into a multitude of trivial calculations.

Plate 3.1 A marine diesel engine main structure being welded. Note the cast steel joints at the ends of the rectangular tubes. (Courtesy: Ruston–Paxman Diesels.)

Plate 3.2 A prefabricated section of hull being lowered into position on the Esso Malaysia. (Courtesy: Esso Petroleum.)

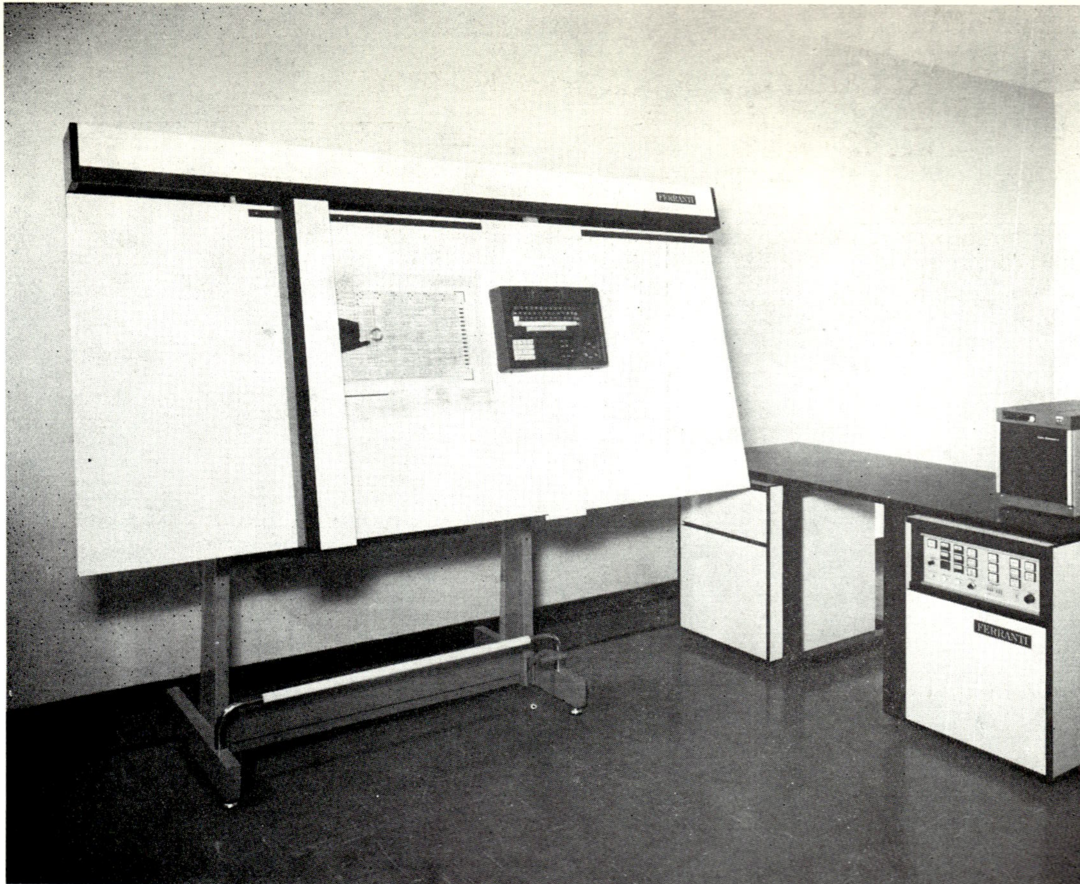

Plate 4.1 A typical drawing digitiser. The unit on the right is necessary for transmitting the information to the

Plate 4.2 A numerically controlled draughting machine used for aircraft lofting. (Courtesy: Hawker Siddeley Aviation Ltd.)

Plate 4.3 An incremental plotter showing the pen carriage and controls (far right). The machine is capable of taking a full-width roll of paper, suitable for engineering drawings, though here it is loaded with a narrow roll. (Courtesy: Computer Instrumentation Ltd.)

Plate 4.4 A graphic display console being used for the isometric drawing of a piping network at the Computer Aided Design Centre, Cambridge. The light pen is in the operator's hand and the eight function keys can be seen below the screen on the left. The alpha-numeric keyboard is in front of the function keys. The teletype on the right is part of another installation.
(Courtesy: GEC-Elliott Automation Ltd.)

Plate 4.5 (a) The general layout of the drilling machine

Plate 4.5 (b) N/C *controls*

Plate 4.5 (c) Drilling heads. These can be controlled to produce all the necessary end-fixing holes in one pass. (Courtesy: Sheepbridge Engineering Ltd.)

(a)

(b)

Plate 5.1 Two examples of a traditional mould loft: (a) is at A.G. Weser (Bremen) and shows the lofting of the Esso London; (b) is the Hawker Siddeley Aviation loft—note the spline and lofting weights. (Courtesy: Esso Petroleum and Hawker Siddeley Aviation.)

5 The Numerical Description of Curves and Surfaces

5.0 WHY DESCRIBE CURVES AND SURFACES IN THIS WAY?

The traditional engineering drawing, with few exceptions, comprises straight lines, and curves generated as arcs of circles. This has a twofold advantage: the lines are simple and quick to produce and the geometry is easily reproduced by standard machining and manufacturing methods. With the advent of numerical control, in particular, continuous path milling and other forms of profiling, the manufacturer has a far greater flexibility to produce complex shapes and curves.

The demand for continuous-path profiling was originally created by those industries where curves had been manufactured by some form of hand-shaping process. Most of them, for example, aircraft, automotive, ship and cam manufacturers, produced shaped components more out of necessity than choice. It is easy to find examples where designers have made a deliberate effort to approximate these complex shapes to simple geometrical forms, e.g. the conventional manufacture of cams on a milling machine, where an inclined dividing head is driven from the table lead screw to generate a smooth curve between specified points, or attempts at eliminating double-curvature skin panels on light aircraft. With the introduction of numerical shape description it is possible not only to satisfy these existing requirements but also to optimise the shape of many other components.

5.1 THE SCOPE OF NUMERICAL DESCRIPTION

It can be seen that, in the past, the wise designer avoided complex shapes for purposes of economic design and manufacture. Even with forming processes such as forging, moulding and casting, the simplicity of making dies and patterns encouraged simple geometrical shapes.

Once the pattern or die has been made, the manufacturing cost of a forming process is normally independent of shape. On the other hand many of these shapes manufactured in the orthodox way are only crude approximations to the desired form.

Consider two examples, the first a connecting rod from a high-performance engine, and the second its near neighbour, a piston. The connecting rod is rather a fine blend of straight lines and arcs of circles. This item is normally forged, but can be cast. Its shape enables a relatively simple die to be sunk, but is it the optimum? The design requirements for a connecting rod are that it should transmit a reciprocating force between the piston and the crankshaft; at the same time the lighter the rod the lower the inertia forces on the system. In the ideal rod the stress throughout would be constant. Although this can never be achieved in practice, a better shape can be generated, reducing stress concentrations and removing redundant material, to obtain a more efficient general stress level. A saving in weight leads to a saving in material. With the piston the problem is a little different. The designer is trying to achieve a manufacturable internal shape capable of dissipating the heat of combustion in such a way that the thermal growth of the piston is fully controlled, whilst maintaining a minimum piston weight to reduce inertia forces and, as a consequence, material costs.

In both these examples, success hinges on the ability of the designer to communicate effectively the shape to the manufacturer. By providing the shape data in a numerical form they can be used directly in the control of a milling or other profiling machine.

Situations occur where the designer is able to predict exactly the profile he requires, but then he approximates it in order to dimension a drawing. This arises in the design of fluid machinery, e.g. blading, volute shapes and valve profiles.

A good example of a situation where numerical techniques have aided design and production is gas-turbine blading. Here, experiments and analyses have enabled the designer to predict the optimum shape of a blade. It was realised that crude approximations to this relatively complex twisted aerofoil form lead to a loss of efficiency. In the early days of gas turbines, manual methods had to be employed to help generate these forms but, more recently, numerical techniques have taken over.

So far the discussion has been limited to shaped surfaces, and no mention has been made of simple curves. It will be seen in the following sections that curves are necessary in defining a shape, but they do have direct uses. The most obvious is the profile of a cam, but another interesting application is in civil engineering, in particular, road layout. Traditionally, the path of a road was defined by circular arcs and straight lines. This becomes very tedious, since the curves must be made to fit through interscetions with other roads and bridging points on rivers, etc. Furthermore, the radii involved preclude the use of a giant compass on site, consequently the site engineer must be provided with calculated values appropriate to his surveying methods. With the introduction of numerical curve fitting, using polynomials, a rapid, smooth fit can be made through a set of defined points whilst providing the necessary data for the site survey. A secondary advantage results from this method in that the driver following a polynomial curve is more alert than when following the arc of a circle with a fixed steering position. An example of this layout is described in detail by Calogero[14].

Summarising:

(i) In general, forming processes such as moulding, casting, forging and profile machining, once a master has been produced, can accommodate an 'ideal' shape at no extra cost.
(ii) The advantage of numerical shape definition is in communicating information between design and manufacture.

5.2 THE TRADITIONAL LOFTING METHODS

There are essentially three steps that are necessary to produce a shaped component, assuming that the curve is not a simple circular arc. First, the shape must be established in some way, possibly by freehand sketching or by producing an engineering drawing with sufficient elevations to establish the general form, or even by defining the mathematical shape equation. Traditionally, the second step has involved drawing a full-size layout of the shape at various sections, from which was cut a nest of templates for checking the profile at stages in manufacture. Finally there is manufacture itself, which may involve hand shaping of the component to match the nested templates, or using the template form as a manufacturing tool as with pressing, stretching and profile-copy milling.

The drawing out of full-size forms has come to be known as lofting. This stems from boat building where, in the past, an upper floor existed above the workshops upon which full-size nested sections of the vessel were drawn using chalk, providing a master drawing against which the boatbuilders could check the profiles of the hull. This came to be known as the mould loft, for obvious reasons, and hence the title lofting.

When a ship's hull, or, for that matter, any profile is lofted, sections of the hull are superimposed one upon the other. The system of nesting is shown in Figure 5.1. Basically only two sets of sections are required, as the third set only duplicates the description. As with engineering drawing, there are several ways in which the sections can be presented and this may vary from company to company.

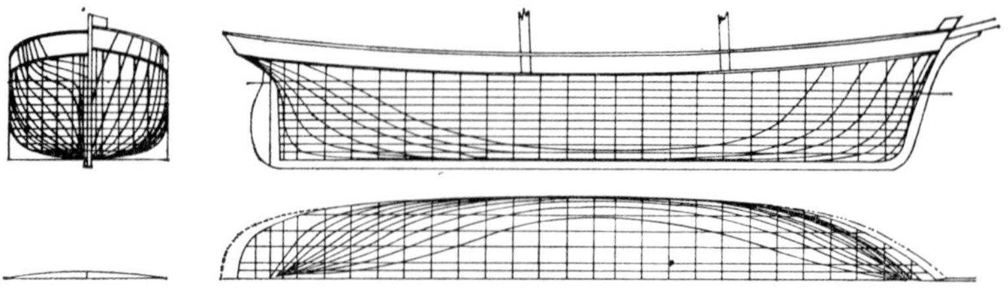

Figure 5.1 The mould loft lines of the schooner 'Mary Ashburner'—from 'The Merchant Schooners' by Basil Greenhill. (Courtesy: David and Charles, Publishers.) Note the advantage taken of symmetry by lofting only half the hull in plan and by showing aft and forward sections side by side

Plate 5.1 shows two traditional forms of mould loft: in one, aircraft loftsmen are seen crawling over lofting tables whilst laying out the aircraft shape in pencil on metal loft plates and, in the other, shipyard template makers carry a nest of wooden templates across the loft; notice the curves drawn on the floor.

Short lengths of curve with relatively high rates of curvature are often drawn with French curves, but for most applications the curves are too long and shallow to make this a practical proposition. The normal procedure is to employ slender strips of wood or plastic known as lofting splines. These are laid on the lofting surface and are bent to shape so that one side of the spline is made to pass through each of the points on the curve; to avoid the spline springing back to its undeformed shape, it is weighted at each point with lofting weights. When the spline is finally in position it can be secured by additional intermediate weights, providing an edge along which the line may be drawn. The arrangement is shown diagramatically in Figure 5.2. Also shown are the equivalent forces acting upon the spline, from which it can be seen that the spline behaves as a continuous slender beam, the point constraints resulting in applied loads and reactions at intervals

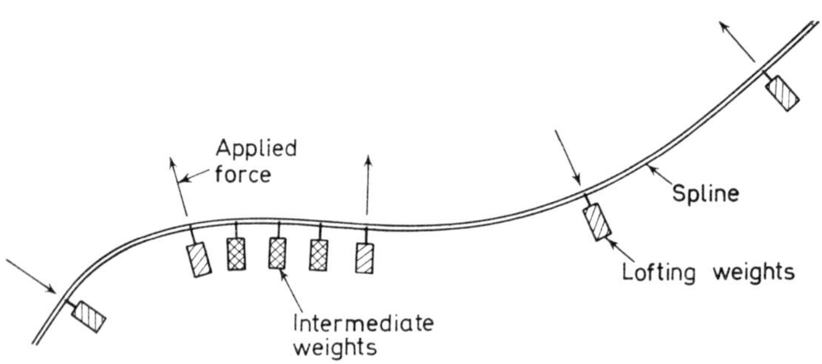

Figure 5.2 Diagrammatic layout of spline and lofting weights. The arrows denote the directions of the reaction forces

along the beam. Note that intermediate weights should not contribute to the bending moment along the beam, but merely steady the spline whilst the line is being drawn.

5.3 THE CUBIC SPLINE

To define a curve numerically it is necessary to calculate the coordinates of closely spaced intermediate points along the curves between those specified. Therefore, an approximate mathematical equation for each section of curve is required; these equations form a piecewise spline. The obvious choice of equation for each section would appear to be a cubic polynomial of the form

$$y = a+bx+cx^2+dx^3$$

where a, b, c and d are constants. The choice seems logical for two reasons. First, to establish the four constants, four known values of x and y must exist; for one section of curve these would be the two end points (A and B in Figure 5.3) to that section plus one further

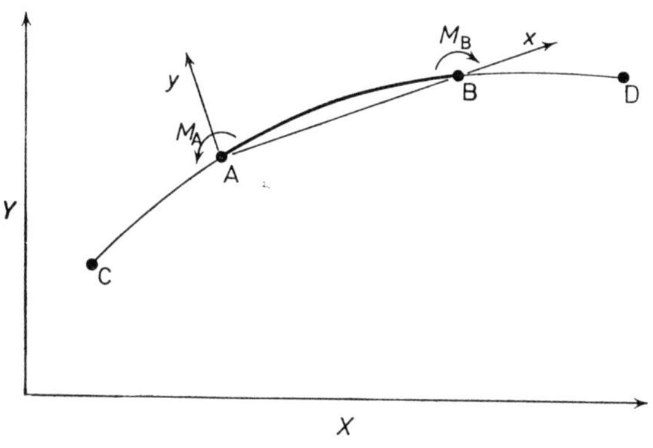

Figure 5.3 The coordinates for bending theory applied to a section of spline between A *and* B, *relative to a standard frame of reference*

point beyond each of these (C and D). An even-order polynomial such as a quadratic or a quartic requires an odd number of values to define the constants destroying the symmetry of the arrangement. The second reason follows from the analogy of the spline with the continuous beam. The basic equation relating the geometry of beam deflection to the applied moment is

$$\frac{d^2y}{dx^2} = \frac{M}{EI} \tag{5.1}$$

where E and I are Young's modulus of elasticity and the second moment of area of the section, respectively; x is the coordinate along the chord joining the points of constraint and y the coordinate normal to the chord.

The bending moment M of the section of spline between points A and B will be a linear function of x. If, therefore, equation (5.1) is integrated twice to obtain a relationship between x and y, the resulting expression will be the classical equation for bending which is an equation in x^3. It would therefore seem sensible to approximate to the spline by using a cubic polynomial, i.e. a separate cubic polynomial for each of the three sections of curve: CA, AB and BD. This has in fact been done. To be useful, cubic polynomials for these three sections should have the same frame of reference, namely a cartesian coordinate system in X and Y as shown in Figure 5.3.

A Considerable amount of work has been carried out on the cubic spline in this form and on improved variations and much of this is reported in reference 15.

The simple cubic spline has a number of drawbacks. The obvious one is the disparity between the frame of reference used for the classical treatment of bending, i.e. the chord joining the constraints, and that used in practice—a standard frame applied to all sections of the curve regardless of their chord directions. It follows that sections having chords almost parallel to the standardised X coordinate will more truly represent the bending of a beam, whilst those with chords almost normal to the x axis will deviate wildly from this representation. In the latter case there is the added disavantage of the first derivative (mathematical tangent) of the curve becoming infinite. Furthermore, it will be recalled from basic beam bending theory that equation (5.1) is based on the assumption that deflections are small compared with the distance between supports. This assumption would only be applicable to very shallow curves.

On the credit side, this analogy with bending theory can be quite useful in establishing a relationship between adjoining sections of curve. It is obvious that if the total curve is to be smooth, the end points of each individual section must coincide with the end point of the adjacent section. Furthermore, the slope of the curve where the two sections join must be maintained, i.e. the first derivative must be continuous, otherwise a kink would occur. The aspect which is not so obvious is that the second derivative must also be continuous along the total length of the curve. This becomes apparent from a study of Figure 5.3. Consider the point A; here the bending moment M_A is common to the sections CA and AB and, if this is the case, equation (5.1) is valid throughout point A, making the second derivative continuous. This is where a curve produced by a spline can differ from that drawn with French curves. With the latter, only the continuity of slope, or first derivative, can be guaranteed, and this only by careful inspection by the user.

Summarising:

(i) The simple cubic spline is not a true representation of the bent beam since their coordinates are based on different axes.

5.4 THE PARAMETRIC SPLINE

To overcome the difficultes encountered with the simple cubic spline, Ferguson[16] devised a parametric cubic spline, the vectorial derivation of which follows.

Consider the four vectors shown in Figure 5.4, the resultant of which is $\boldsymbol{R}$. If each vector is multiplied by some scalar quantity U the effect will be to multiply the resultant $\boldsymbol{R}$ by U. As the scalar U is

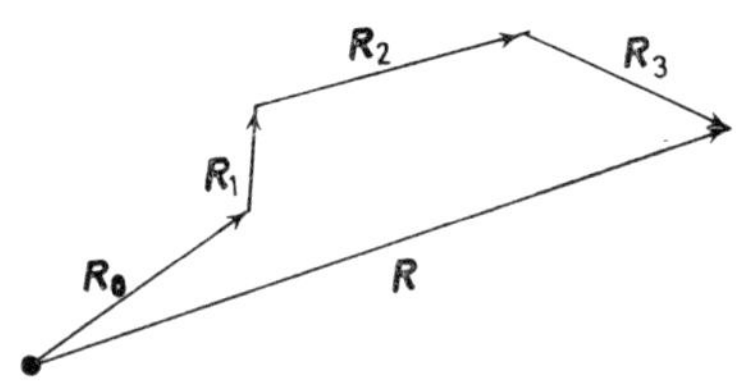

Figure 5.4 Vector $\boldsymbol{R}$ *is the resultant of* $\boldsymbol{R}_0+\boldsymbol{R}_1+\boldsymbol{R}_2+\boldsymbol{R}_3$

increased, $\boldsymbol{R}$ increases in length and its direction is unchanged, i.e. the end of $\boldsymbol{R}$ traces out a straight line. If, however, each of the four vectors $\boldsymbol{R}_0\ldots\boldsymbol{R}_3$ is multiplied by a different scalar, then not only is the length of $\boldsymbol{R}$ increased but its direction is changed; the end of $\boldsymbol{R}$ now traces out a curve.

Ferguson defines the spline as the path traced out by the tip of the vector $\boldsymbol{R}(U)$. This vector is related to the parameter U by the following equation:

$$\boldsymbol{R}(U) = \boldsymbol{R}_0+\boldsymbol{R}_1U+\boldsymbol{R}_2U^2+\boldsymbol{R}_3U^3 = \sum_{i=0}^{i=3} \boldsymbol{R}_iU^i \tag{5.2}$$

Consider the single section of spline in Figure 5.5. The positions of the two end points are given by the vectors $\boldsymbol{A}$ and $\boldsymbol{B}$. In addition the tangents at the points $\boldsymbol{A}$ and $\boldsymbol{B}$ are defined by the slope vectors $\boldsymbol{T}_A$ and $\boldsymbol{T}_B$. If the value of U in equation (5.2) is allowed to vary between 0 at $\boldsymbol{A}$ and 1 at $\boldsymbol{B}$ then,

$$\boldsymbol{R}(0) = \boldsymbol{A} \qquad \boldsymbol{R}(1) = \boldsymbol{B}$$

$$\left.\frac{\mathrm{d}\boldsymbol{R}}{\mathrm{d}U}\right|_{U=0} = \boldsymbol{T}_A \qquad \text{and} \qquad \left.\frac{\mathrm{d}\boldsymbol{R}}{\mathrm{d}U}\right|_{U=1} = \boldsymbol{T}_B$$

In equation (5.2) there are four unknown vectors: $\boldsymbol{R}_0$, $\boldsymbol{R}_1$, $\boldsymbol{R}_2$, and $\boldsymbol{R}_3$. Substituting in the end point conditions (ie $U = 0$ or 1) gives

$$\boldsymbol{A} = \boldsymbol{R}_0 \qquad \text{and} \qquad \boldsymbol{B} = \boldsymbol{R}_0+\boldsymbol{R}_1+\boldsymbol{R}_2+\boldsymbol{R}_3$$

Differentiating equation (5.2) and substituting gives

$$\boldsymbol{T}_A = \boldsymbol{R}_1 \qquad \text{and} \qquad \boldsymbol{T}_B = \boldsymbol{R}_1+\boldsymbol{R}2\boldsymbol{R}_2+3\boldsymbol{R}_3$$

Eliminating the unknowns between these four equations yields:

$$\boldsymbol{R}(U) = U^3[2(\boldsymbol{A}-\boldsymbol{B})+\boldsymbol{T}_A+\boldsymbol{T}_B]+U^2[3(\boldsymbol{B}-\boldsymbol{A})-2\boldsymbol{T}_A-\boldsymbol{T}_B]+U\boldsymbol{T}_A+\boldsymbol{A} \tag{5.3}$$

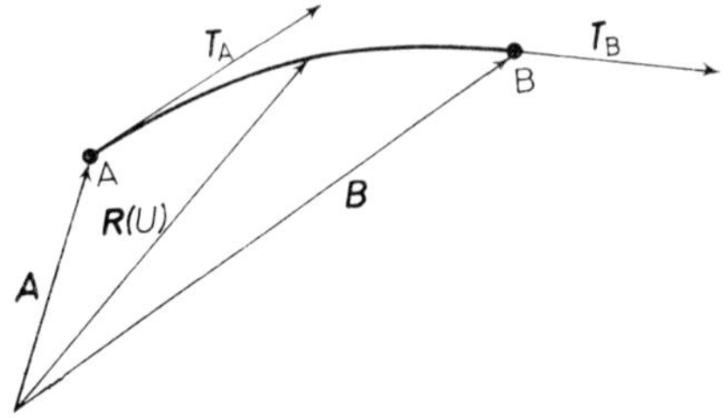

Figure 5.5 A section of curve between two points A *and* B

So far only one section of curve between two given points has been considered. In practice the designer is normally concerned with fitting a spline through several data points. Clearly the result just derived may be applied to the intervals between adjoining points, but in a real situation the only available information about the tangents would be at the starting and finishing positions of the entire curve, where the shape would be required to match up to adjoining surfaces. It is necessary, therefore, to establish a means of arriving at values for the

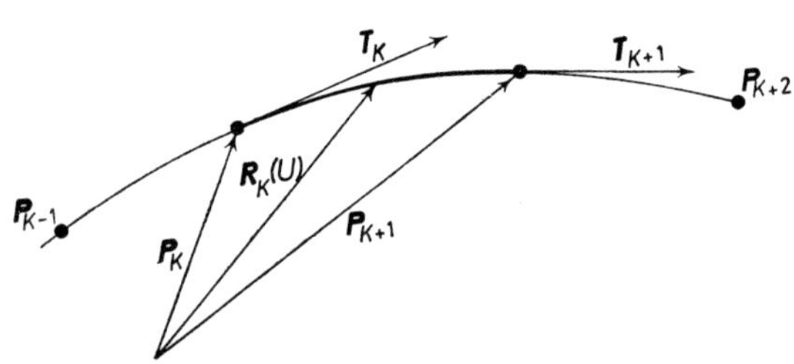

Figure 5.6 A general section of the spline between points $\boldsymbol{P}_K$ and $\boldsymbol{P}_{K+1}$ defined by $\boldsymbol{R}_K(U)$

slope vectors at the intermediate points. Figure 5.6 shows a general section of the spline spanning several given points. Consider any general point $\boldsymbol{P}_K$ on the curve then, taking the section of curve $\boldsymbol{R}_K(U)$ between $\boldsymbol{P}_K$ and $\boldsymbol{P}_{K+1}$, gives by analogy with equation (5.3):

$$\begin{aligned}\boldsymbol{R}_K(U) = {} & U^3[2(\boldsymbol{P}_K-\boldsymbol{P}_{K+1})+\boldsymbol{T}_K+\boldsymbol{T}_{K+1}] \\ & +U^2[3(\boldsymbol{P}_{K+1}-\boldsymbol{P}_K)-2\boldsymbol{T}_K-\boldsymbol{T}_{K+1}]+U\boldsymbol{T}_K+\boldsymbol{P}_K \end{aligned} \tag{5.4}$$

In this equation the values of $\boldsymbol{P}_K$ and $\boldsymbol{P}_{K+1}$ are known, and U is the parameter which is to be varied between 0 and 1 to establish intermediate values of $\boldsymbol{R}_K(U)$. The problem is now one of finding a relationship to establish values for the two unknowns in the equation, i.e. the two slope vectors.

It is known that the spline is continuous since adjacent sections of the curve share the same end points, and it can also be appreciated that to avoid kinks the slope must be continuous. From the analogy with the bent beam it was shown that a continuous bending moment implies a continuous second derivative. This final condition of continuous second derivative is now applied.

Differentiating equation (5.4) twice with respect to U:

$$\boldsymbol{R}''_K(U)=6U[2(\boldsymbol{P}_K-\boldsymbol{P}_{K+1})+\boldsymbol{T}_K+\boldsymbol{T}_{K+1}]+2[3(\boldsymbol{P}_{K+1}-\boldsymbol{P}_K)-2\boldsymbol{T}_K-\boldsymbol{T}_{K+1}] \tag{5.5}$$

By a direct comparison with equation (5.5) an expression can be found for the second derivative of the curve between $\boldsymbol{P}_{K+1}$ and $\boldsymbol{P}_{K+2}$ which is

$$\begin{aligned}\boldsymbol{R}''_{K+1}(U) = {} & 6U[2(\boldsymbol{P}_{K+1}-\boldsymbol{P}_{K+2})+\boldsymbol{T}_{K+1} \\ & +\boldsymbol{T}_{K+2}]+2[3(\boldsymbol{P}_{K+2}-\boldsymbol{P}_{K+1})-2\boldsymbol{T}_{K+1}-\boldsymbol{T}_{K+2}] \end{aligned} \tag{5.6}$$

If the second derivative (curvature) is to be continuous between the two adjoining sections, then it must have the same value just to either side of the point $\boldsymbol{P}_{K+1}$; this may be written mathematically as

$$\boldsymbol{R}''_K\,(1) = \boldsymbol{R}''_{K+1}\,(0)$$

giving, from equations (5.5) and (5.6),

$$\begin{aligned}6[2(\boldsymbol{P}_K-\boldsymbol{P}_{K+1})+\boldsymbol{T}_K+\boldsymbol{T}_{K+1}]+2[3(\boldsymbol{P}_{K+1}-\boldsymbol{P}_K)-2\boldsymbol{T}_K-\boldsymbol{T}_K-\boldsymbol{T}_{K+1}] \\ = 2[3(\boldsymbol{P}_{K+2}-\boldsymbol{P}_{K+1})-2\boldsymbol{T}_{K+1}-\boldsymbol{T}_{K+2}]\end{aligned}$$

Rearranging and simplifying:

$$\boldsymbol{T}_K+4\boldsymbol{T}_{K+1}+\boldsymbol{T}_{K+2} = 3(\boldsymbol{P}_{K+2}-\boldsymbol{P}_K) \tag{5.7}$$

For the points $\boldsymbol{P}_0$, $\boldsymbol{P}_1$, ... $\boldsymbol{P}_K$...$\boldsymbol{P}_N$ equation (5.7) yields a set of N-1 equations. Using elementary matrix notation these equations can be better represented in the following form:

$$\begin{bmatrix} 4 & 1 & 0 & 0 & 0 \\ 1 & 4 & 1 & 0 & 0 \\ & 1 & 4 & 1 & \\ 0 & & 1 & 4 & 1 \\ 0 & 0 & & 1 & 4 \end{bmatrix} \begin{bmatrix} \boldsymbol{T}_1 \\ \boldsymbol{T}_2 \\ \boldsymbol{T}_K \\ \boldsymbol{T}_{N-2} \\ \boldsymbol{T}_{N-1} \end{bmatrix} = \begin{bmatrix} 3(\boldsymbol{P}_2-\boldsymbol{P}_0)-\boldsymbol{T}_0 \\ 3(\boldsymbol{P}_3-\boldsymbol{P}_1) \\ 3(\boldsymbol{P}_{K+1}-(\boldsymbol{P}_{K-1}) \\ 3(\boldsymbol{P}_{N-1}-\boldsymbol{P}_{N-3}) \\ 3(\boldsymbol{P}_N-\boldsymbol{P}_{N-2})-\boldsymbol{T}_N \end{bmatrix} \tag{5.8}$$

The unknown slope vectors are in the column matrix (it is assumed that the two tangents $\boldsymbol{T}_0$ and $\boldsymbol{T}_N$ are known). The matrix equation (5.8) is in a vector and not a scalar form. To solve it, it is necessary to realise that a three-dimensional vector has three components, in the x, y and z directions (unit vectors $\boldsymbol{i}, \boldsymbol{j}$ and $\boldsymbol{k}$). This means that all the x components must be equated and similarly the y and z components. In other words, the three directions must be equated and satisfied one at a time.

Summarising:

(i) The relations between the tangent vectors along a parametric cubic spline are derived from the continuity of curvature along the spline.

5.5 EXAMPLES AND EXERCISES IN THE USE OF THE PARAMETRIC SPLINE

The following examples have a threefold purpose. The first is to enable the reader to relate the analysis to real problems; as a consequence it will be seen that, although the analysis was devised to handle complex problems on the digital computer, it is still a useful tool for a designer–draughtsman with a desk calculator, enabling him to specify simple curves. Finally it serves to show that even the best pieces of analysis have their limitations.

EXAMPLE 1

A glass envelope is required for the structure of a high-power electric light bulb. The profile of the bulb has been specified by the following five points:

x	y
0	60
150	92
250	138
350	62
365	0

The dimensions are in millimetres and x is the coordinate along the centreline of the bulb, starting at the cap; y is the magnitude of the radius at each of these positions. In order that a forming tool can be made, a smooth curve must be fitted through these points. Having established the equations for this curve it will be assumed that five intermediate points between each pair are adequate to plot the general profile for a visual inspection.

Solution

No end-point slope vectors have been specified, so these must now be established. At the end of the bulb away from the cap the slope vector will have only one component in a radial direction. If small blending radii are neglected the end of the curve nearest the cap will have only an axial component. To start with, these will be taken as:

$$\boldsymbol{T}_0 = 10\boldsymbol{i}+0\boldsymbol{j}$$

and

$$\boldsymbol{T}_N = 0\boldsymbol{i}-10\boldsymbol{j}$$

where $\boldsymbol{i}$ and $\boldsymbol{j}$ are unit vectors in the x and y directions respectively. (Note the negative sign in $\boldsymbol{T}_N$; this is in order that the directions of the tangent vectors follow around the curve.)

The matrix equation (5.8) is made up of three-dimensional vectors but, since the envelope in question is a solid of revolution, it becomes a two-dimensional problem in x and y only. Substituting the x components into equation (5.8):

$$\begin{bmatrix} 4 & 1 & 0 \\ 1 & 4 & 1 \\ 0 & 1 & 4 \end{bmatrix} \begin{bmatrix} \boldsymbol{T}_{x1} \\ \boldsymbol{T}_{x2} \\ \boldsymbol{T}_{x3} \end{bmatrix} = \begin{bmatrix} 3(250-0)-10 \\ 3(350-150) \\ 3(365-250)-0 \end{bmatrix}$$

i.e.

$$4\boldsymbol{T}_{x1}+\boldsymbol{T}_{x2} = 740 \qquad \text{(i)}$$

$$\boldsymbol{T}_{x1}+4\boldsymbol{T}_{x2}+\boldsymbol{T}_{x3} = 600 \qquad \text{(ii)}$$

$$\boldsymbol{T}_{x2}+4\boldsymbol{T}_{x3} = 345 \qquad \text{(iii)}$$

Eliminating between the equations:

$$\text{(ii)} -\tfrac{1}{4}\times \text{(iii)} \quad \boldsymbol{T}_{x1}+4\boldsymbol{T}_{x2}-\tfrac{1}{4}\boldsymbol{T}_{x2}+\boldsymbol{T}_{x3}-\boldsymbol{T}_{x3} = 600-\tfrac{345}{4}$$

$$\boldsymbol{T}_{x1}+3{\cdot}75\boldsymbol{T}_{x2} = 513{\cdot}75 \qquad \text{(iv)}$$

$$(\text{i})-\tfrac{1}{3.75}\times\ (\text{iv})\qquad 4\boldsymbol{T}_{x1}-\frac{\boldsymbol{T}_{x1}}{3{\cdot}75}+\boldsymbol{T}_{x2}-\boldsymbol{T}_{x2}=740-\tfrac{513.75}{3.75}$$

$$3{\cdot}733\boldsymbol{T}_{x1}=603 \qquad (\text{v})$$

from (v) $\quad \boldsymbol{T}_{x1}=161{\cdot}5$
from (iv) $\quad \boldsymbol{T}_{x2}=93{\cdot}9$
from (iii) $\quad \boldsymbol{T}_{x3}=62{\cdot}8$

Substituting the y components into equation (5.8):

$$\begin{bmatrix}4 & 1 & 0\\ 1 & 4 & 1\\ 0 & 1 & 4\end{bmatrix}\begin{bmatrix}\boldsymbol{T}_{y1}\\ \boldsymbol{T}_{y2}\\ \boldsymbol{T}_{y3}\end{bmatrix}=\begin{bmatrix}3(138-60)-0\\ 3(62-92)\\ 3(0-138)+10\end{bmatrix}$$

This may be solved by elimination as before, giving the values in the following table.

Point	*Coordinates* x	y	*Slope-vector components* $\boldsymbol{T}_x$	$\boldsymbol{T}_y$
0	0	60	10	0
1	150	92	161·5	61·8
2	250	138	93·9	−13·5
3	350	62	62·8	−97·6
4	365	0	0	−10

The next step is to consider a length of spline between two of the specified points, e.g. the length between points $K=0$ and $K=1$. For this section, equation (5.4) becomes:

$$\boldsymbol{R}_0(U)=U^3[2(\boldsymbol{P}_0-\boldsymbol{P}_1)+\boldsymbol{T}_0+\boldsymbol{T}_1]+U^2[3(\boldsymbol{P}_1-\boldsymbol{P}_0)-2\boldsymbol{T}_0-\boldsymbol{T}_1]+U\boldsymbol{T}_0+\boldsymbol{P}_0 \qquad (\text{vi})$$

Since five equispaced intermediate points are required, each section of spline will be divided into six. The increments in U between 0 and 1 will be $\frac{1}{6}$. As equation (vi) is in vector form the individual components must be equated. For the components in the x direction it will be

$$\boldsymbol{R}_{x0}(U)=U^3[2(\boldsymbol{P}_{x0}-\boldsymbol{P}_{x1})+\boldsymbol{T}_{x0}+\boldsymbol{T}_{x1}]$$
$$+U^2[3(\boldsymbol{P}_{x1}-\boldsymbol{P}_{x0})-2\boldsymbol{T}_{x0}-\boldsymbol{T}_{x1}]+U\boldsymbol{T}_{x0}+\boldsymbol{P}_{x0}$$

For the first intermediate point this becomes

$$\boldsymbol{R}_{x0}\left(\frac{1}{6}\right)=\frac{1}{6^3}[2(0-150)+10+161{\cdot}5]+\frac{1}{6^2}[3(150-0)-2\times10-161{\cdot}5]$$
$$+\frac{1}{6}\times10+0\text{ mm}$$
$$=8{\cdot}5\text{ mm}$$

by taking increments in U of $\frac{1}{6}$ over the first section of spline, the following values of the x component result:

$$\begin{array}{llllllll} U = 0 & \frac{1}{6} & \frac{1}{3} & \frac{1}{2} & \frac{2}{3} & \frac{5}{6} & 1 \\ x = 0 & 8{\cdot}5 & 28{\cdot}4 & 56 & 87{\cdot}9 & 120{\cdot}4 & 150 \end{array}$$

The same exercise can be repeated for the y component. In *Table* (5.1) the coordinates of the points along the curve are listed. These values are plotted in Figure 5.7, providing a fair representation of the desired shape.

Table 5.1

x	y	x	y
0	60	250	138
8	60	266	133
28	63	285	123
56	68	303	110
87	74	321	95
120	82	337	78
150	92	350	62
175	102	358	46
192	114	363	32
208	124	365	20
222	132	365	10
235	137	365	3
		365	0

Note that these values are rounded to the nearest integer and as a consequence small variations towards the end of the table are not apparent.

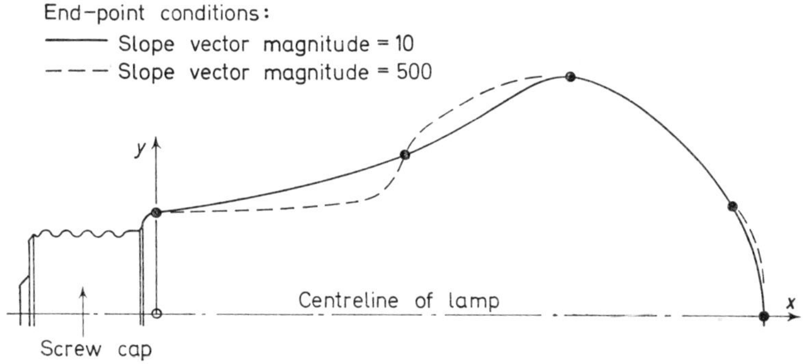

Figure 5.7 Solid of revolution for electric lamp. Note the effect of taking a large magnitude for the end-point slope vectors

The directions of the slope vectors at the end positions are determined by the slope of the connecting curve, but their magnitudes are undecided. In this example a magnitude of 10 was selected which produced a fair curve. Had an extremely large value been used the blend at the end point would have approached a straight line, which in some situations may be desirable, but the general curve may have been forced away from a smooth spline resulting in a wavy contour passing through the given points, as shown in Figure 5.7 for slope vectors having a magnitude of 500. Selection of the magnitudes of the end-point vectors is a compromise between providing a fair curve and applying sufficient constraint to ensure matching with other curves.

EXAMPLE 2

Consider the section of curve shown in Figure 5.8(a). The slope vectors selected for the section are very large compared with the chord length.

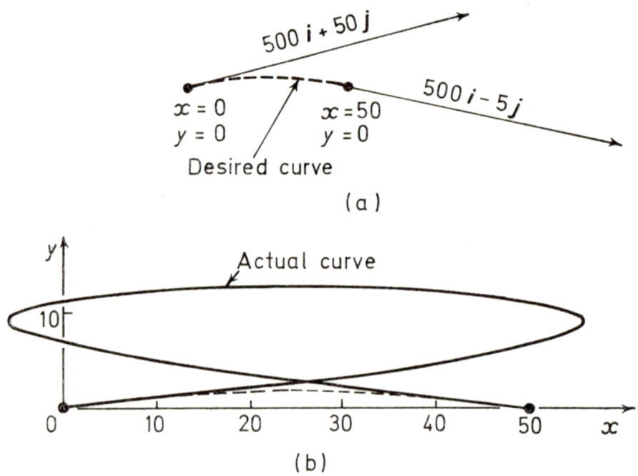

Figure 5.8 The effect of extra-large slope vectors on a section of spline. (a) The expected curve; (b) The actual curve resulting from the parametric cubic spline

Assuming the curve to run between points 0 and 1, equation (5.4) becomes equation (vi) in the previous example. The shape of the spline may be calculated in a tabular form as follows.

For the x components:

$$2(\boldsymbol{P}_{x0}-\boldsymbol{P}_{x1})+\ \boldsymbol{T}_{x0}+\boldsymbol{T}_{x1} = 2(0-50)+500+500 = 900$$

$$3(\boldsymbol{P}_{x1}-\boldsymbol{P}_{x0})-2\boldsymbol{T}_{x0}-\boldsymbol{T}_{x1} = 3(50-0)-1000-500 = 1350$$

$$\boldsymbol{T}_{x0} = 500 \qquad \boldsymbol{P}_{x0} = 0$$

and for the y components:

$$2(\boldsymbol{P}_{y0}-\boldsymbol{P}_{y1})+\boldsymbol{T}_{y0}+\boldsymbol{T}_{y1} = 2(0-0)+50-50 = 0$$
$$3(\boldsymbol{P}_{y1}-\boldsymbol{P}_{y0})-2\boldsymbol{T}_{y0}-\boldsymbol{T}_{y1} = 3(0-0)-100+50 = -50$$
$$\boldsymbol{T}_{y0} = 50 \qquad \boldsymbol{P}_{y0} = 0$$

U	U^2	U^3	$900\ U^3$	$-1350\ U^3$
0	0	0	0	0
0·1	0·01	0·001	0·9	−13·5
0·2	0·04	0·008	7·2	−52
0·3	0·09	0·027	24·3	−121·5
0·4	0·16	0·064	57·6	−216
0·5	0·25	0·125	112·5	−337·8
0·6	0·36	0·216	194·4	−486
0·7	0·49	0·343	308·7	−662
0·8	0·64	0·512	460·8	−865
0·9	0·81	0·729	656·1	−1093
1·0	1·0	1·0	900	−1350

(*Continued*)

$500\ U$	$R_{x0}(U)$	$-50\ U^2$	$-50\ U$	$R_{y0}(U)$
0	0	0	0	0
50	37·4	−0·5	5	4·5
100	55·2	− 2	10	8
150	52·8	− 4·5	15	10·5
200	41·6	− 8	20	12
250	24·7	−12·5	25	12·5
300	8·4	−18	30	12
350	−3·3	−24·5	35	10·5
400	−4·2	−32	40	8
450	13·1	−40·5	45	4·5
500	50	−50	50	0

The resulting curve is shown in Figure 5.8(b). What would appear to be the sensible curve is shown as a broken line. The loop results from the over-constraint of the vectors impressing their influence on the intermediate curve. As the magnitudes of the vectors approach infinity the ends of the curve are 'forced' into a straight line. The reader may wish to repeat the exercise using smaller vectors. The transition from the sensible curve to the loop should occur at a vector magnitude at about half that given.

QUESTION 1

A linear cam is required to control the mixture ratio of two chemicals in a chemical-plant process. Six experiments have been performed to find the ratios for given mixtures, these provide reference points for the cam profile. They are:

x	0	20	40	60	80	100
y	30	29	18	12	4	0

where x is along the axis of motion of the cam and y is normal to that axis. Determine a fair profile for the cam satisfying the experimental values.

Answer

As a smooth fit is more important than matching the ends of the curve end-point slope vectors having a relatively small magnitude are called for. Try repeating the exercise using large magnitudes.

Assuming identical end-point slope vectors of $(5\boldsymbol{i}+0\boldsymbol{j})$ the intermediate slope vectors are: $(23{\cdot}9\boldsymbol{i}-6{\cdot}61\boldsymbol{j})$, $(19{\cdot}2\boldsymbol{i}-9{\cdot}53\boldsymbol{j})$, $(19{\cdot}2\boldsymbol{i}-6{\cdot}25\boldsymbol{j})$ and $(23{\cdot}9\boldsymbol{i}-7{\cdot}43\boldsymbol{j})$. With end-point slope vectors of $(4\boldsymbol{i}+0\boldsymbol{j})$ they are: $(25\boldsymbol{i}-6{\cdot}61\boldsymbol{j})$, $(19\boldsymbol{i}-9{\cdot}53\boldsymbol{j})$, $(19\boldsymbol{i}-6{\cdot}25\boldsymbol{j})$, and $(25\boldsymbol{i}-7{\cdot}43\boldsymbol{j})$.

QUESTION 2

The pipe of an electrically powered horn is to be manufactured from plastic. It starts as a parallel tube, 20 mm in diameter at the power unit and diverges to a bell mouth at the other end. The die for moulding, the inner profile is to be turned on a profiling lathe, working from a master template. The template maker requires a series of closely spaced points from which to form the curve. The only points provided by the pipe designer are the following:

x	0	90	160	190	200
y	10	13	17	20	30

where x is the axial distance in millimetres from the power unit and y is the radial position of the inner surface from the centreline of the pipe.

Answer

As the end near the power unit is parallel, a large vector parallel to the pipe axis has been selected. At the other end the pipe will terminate in a lip and it is more important that a fair curve be fitted through the points, hence a short vector normal to the axis is chosen. The selected end-point vectors were $(100\boldsymbol{i}+0\boldsymbol{j})$ and $(0\boldsymbol{i}+5\boldsymbol{j})$, giving intermediate slope vectors of $(82{\cdot}5\boldsymbol{i}+4{\cdot}73\boldsymbol{j})$, $(50\boldsymbol{i}+2{\cdot}07\boldsymbol{j})$, $(17{\cdot}5\boldsymbol{i}+7{\cdot}982\boldsymbol{j})$.

QUESTION 3

A conveyor rail passes through a workshop as shown in Figure 5.9. It follows the winding path to avoid other equipment in the workshop and to align with the doors. The conveyor forms part of a paint-spraying process. At A the items are lowered into a cleaning tank and

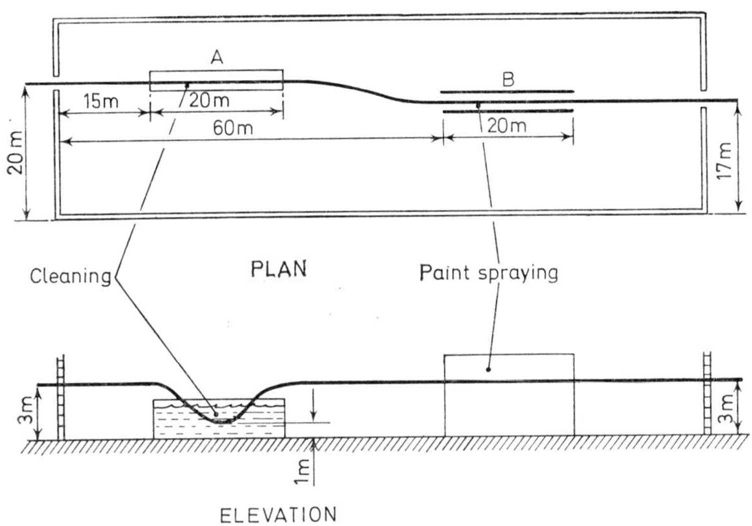

Figure 5.9 The path of a paint-spraying conveyor though a workshop (not to scale)

at B they pass through the spraying booth. The critical dimensions for the track are given in the illustration. Specify a smooth curve for the conveyor to follow, calculating the positions of intermediate points which may be used for location purposes. Plot this curve, exaggerating the scales normal to the direction of motion. Vary the tangent vectors at the end points and find the effects on the curve.

Answer

If the end points are taken to be at the entry to the cleaning tank and at the exit from the spray booth, with slope vector $(16\boldsymbol{i}+0\boldsymbol{j}+0\boldsymbol{k})$, then the intermediate slope vectors become $(6{\cdot}411\boldsymbol{i}-0{\cdot}161\boldsymbol{j}-429\boldsymbol{k})$, $(18{\cdot}357\boldsymbol{i}+0{\cdot}643\boldsymbol{j}+1{\cdot}714\boldsymbol{k})$ and $(25{\cdot}161\boldsymbol{i}-2{\cdot}411\boldsymbol{j}-0{\cdot}429\boldsymbol{k})$.

This curve is shown, together with others taken for various end-point slope vectors, in Figure 5.10.

QUESTION 4

Test the parametric spline against known shapes such as a circle. Start by using widely spaced points along the curve for the known points of the spline. Repeat the exercise with more closely spaced points.

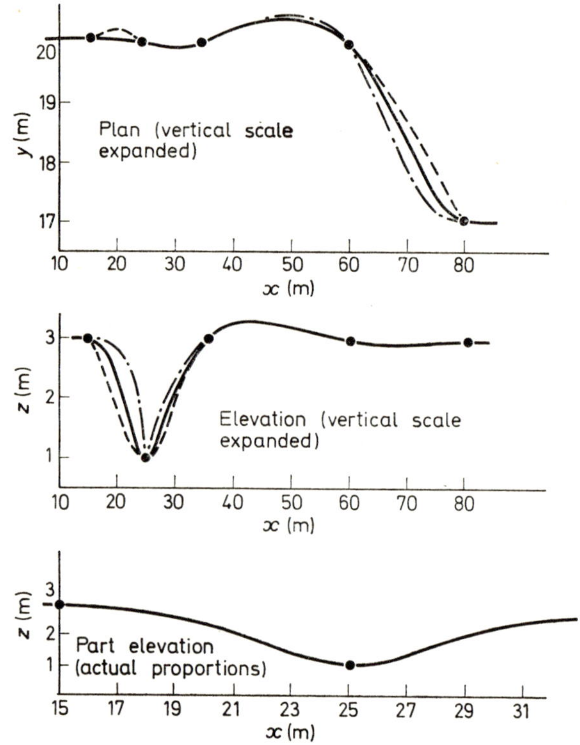

Figure 5.10 The route of the conveyor rail. Although the plan and elevation shown in the upper part of the figure, with expanded scales to illustrate the form of the curves, suggest unreal curves, the part elevation shows the relatively steady curvature. Note in the middle diagram how the curve overshoots the horizontal; in this particular application this may be an advantage

——— $\boldsymbol{T}_0 = \boldsymbol{T}_N = 16\boldsymbol{i}$; - - - $\boldsymbol{T}_0 = \boldsymbol{T}_N = \boldsymbol{i}$;
· —· —· $\boldsymbol{T}_0 = \boldsymbol{T}_N = 32\boldsymbol{i}$

Answer

It should be found that, as the chord length between the specified points becomes small compared with the radius of curvature, the spline will more truly represent the curve.

5.6 SURFACE DEFINITION

One requirement in defining a surface is that the form of the surface can be easily modified. Two desirable properties result from this, the first is that any point on the surface has a unique 'address', and secondly any modification of the surface will not destroy the basic frame of reference.

The storing of a surface within a computer produces a compromise between accuracy and cost; cost being space. The simplest method of definition is by an array of discrete points distributed in close proximity over the surface. If the points are densely distributed, then a fair representation of the surface may be obtained by joining the points with straight lines. To define anything but the simplest surface in this manner would require so many points that the system would be unworkable. The next stage in improvement might be to store points which are more widely spaced, and to join them, or interpolate between them, by means of curves.

A method of surface definition akin to this concept was developed by Coons[17]. In Coons' system, the surface area enclosed by the lines joining four adjacent points is known as a surface patch. When the surface geometry is simple it may be represented by a single patch and,

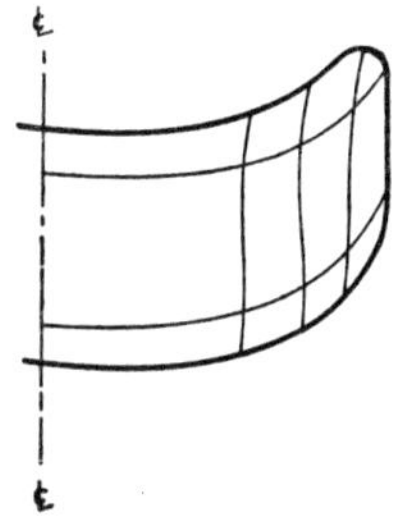

Figure 5.11 An example of how the left-hand half of a motor car windscreen might be divided into surface patches. The areas where complex curvatures exist are represented by several smaller patches whilst the major vision surface may be represented by a single patch

as the surface is modified resulting in greater complexity, the patch may be subdivided into a series of smaller patches to increase the accuracy of definition, e.g. a major part of a car roof may be represented by a single patch, whereas the shape of a windscreen may need several such patches to define it accurately as shown in Figure 5.11. The surface patch is defined by Coons in terms of the conditions existing at its corner points; these are related to the surface by means of a surface equation.

Consider the surface patch shown in Figure 5.12 bounded by the four curves A, B, C and D.

The positions of the corner points are given by the vectors $\boldsymbol{P}_1$, $\boldsymbol{P}_2$, $\boldsymbol{P}_3$

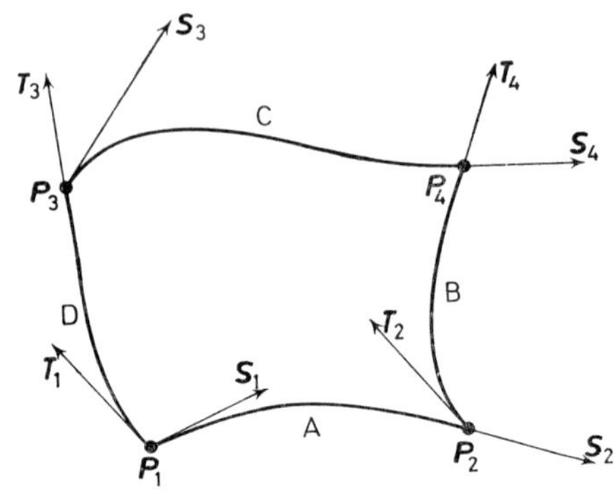

Figure 5.12 A single patch showing the vectors defining the properties of the corner points

and $\boldsymbol{P}_4$ and the properties of the bounding curves at the corners are defined by the slope vectors $\boldsymbol{T}_1$, $\boldsymbol{S}_1$, $\boldsymbol{T}_2$, $\boldsymbol{S}_2$, $\boldsymbol{T}_3$, $\boldsymbol{S}_3$, $\boldsymbol{T}_4$ and $\boldsymbol{S}_4$.

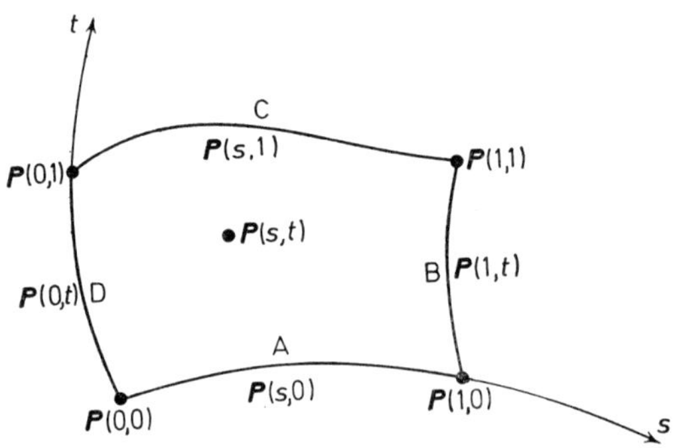

Figure 5.13 The sides and corner points to the patch defined in terms of the parameters s and t. Sides D *and* A *may be considered as axes t and s to a set of curvilinear coordinates as shown*

Consider a general point anywhere on the surface, defined by a vector $\boldsymbol{P}$, and allow the vector $\boldsymbol{P}$ to be a function of two parameters s, and t, where s and t can take on any value between 0 and 1 representing the boundaries of the patch. Therefore

$$\boldsymbol{P} = \boldsymbol{P}(s, t)$$

The corner points now become

$$\boldsymbol{P}_1 = \boldsymbol{P}(0, 0) \qquad \boldsymbol{P}_2 = \boldsymbol{P}(1, 0)$$
$$\boldsymbol{P}_3 = \boldsymbol{P}(0, 1) \qquad \boldsymbol{P}_4 = \boldsymbol{P}(1, 1)$$

as shown in Figure 5.12. Starting at $\boldsymbol{P}_1$ gives $s = t = 0$; if t is held at 0, and s is increased, side A is traced until $\boldsymbol{P}_2$ is reached, when s becomes 1.

If t is increased a small amount and held at this value and the above procedure repeated, a curve lying on the surface and alongside A is created. It now becomes clear that by manipulating the parameters s and t it is possible to reproduce the surface by a mesh of lines spaced at controlled intervals.

The parametric equation for $\boldsymbol{P}(s, t)$ is written in terms of the blended sum of the four boundaries and the four corner points as follows:

$$\begin{aligned}\boldsymbol{P}(s, t) = {} & \boldsymbol{P}(0, t)\, B_0(s) + \boldsymbol{P}(1, t)\, B_1(s) + \boldsymbol{P}(s, 0)\, B_0(t) + \boldsymbol{P}(s, 1)\, B_1(t) \\ & - \boldsymbol{P}(0, 0)\, B_0(s)\, B_0(t) - \boldsymbol{P}(1, 0)\, B_1(s)\, B_0(t) \\ & - \boldsymbol{P}(0, 1)\, B_0(s)\, B_1(t) - \boldsymbol{P}(1, 1)\, B_1(s)\, B_1(t) \end{aligned} \tag{5.9}$$

where B_0 and B_1 are scalar functions known as blending functions.

It should be noted from the first four terms of equation (5.9) that the blending functions are associated with the sides of the patch, i.e. B_0 is associated with the sides $s = 0$ and $t = 0$ and B_1 is associated with the sides $s = 1$ and $t = 1$.

Since the second four terms contain corner points, which are in effect the intersection points of sides, it will be seen that these terms contain

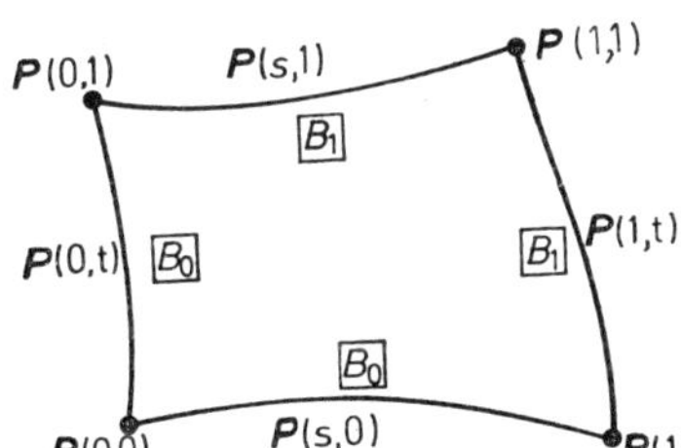

Figure 5.14 The association of blending functions with the sides and corner points

blending functions associated with the two adjacent sides. Figure 5.14 indicates this association diagramatically.

Furthermore, the blending function is always a function of the parameter whose axis is normal to the side it represents, e.g. the axis of s is normal to the side $\boldsymbol{P}(0, t)$, therefore the first term in equation (5.9) becomes

$$\boldsymbol{P}(0, t)\, B_0(s)$$

The significance of these observations is that, moving across the surface from the side $\boldsymbol{P}(0, t)$ to $\boldsymbol{P}(1, t)$, the influence of side $\boldsymbol{P}(0, t)$ on the form of th surface decreases and that of $\boldsymbol{P}(1, t)$ increases, until the boundary is reached where $\boldsymbol{P}(0, t)$ has no influence, and the surface is represented by $\boldsymbol{P}(1, t)$. Hence the title 'blending function'.

The blending functions have not yet been specified, but these functions must satisfy certain conditions:

(a) The functions must be continuous over the region of the patch.
(b) The boundary curves and, as a result, the corner points, must be contained in the surface defined by $\boldsymbol{P}(s, t)$.
(c) Any point on the surface must be uniquely defined by the components of the vector $\boldsymbol{P}(s, t)$ in the x, y and z directions.

Two such functions satisfy these requirements. Consider

$$\left.\begin{aligned} B_0(s) &= 1-3s^2+2s^3 \\ \text{and} \qquad B_1(s) &= 1-B_0(s) = 3s^2-2s^3 \end{aligned}\right\} \qquad (5.10)$$

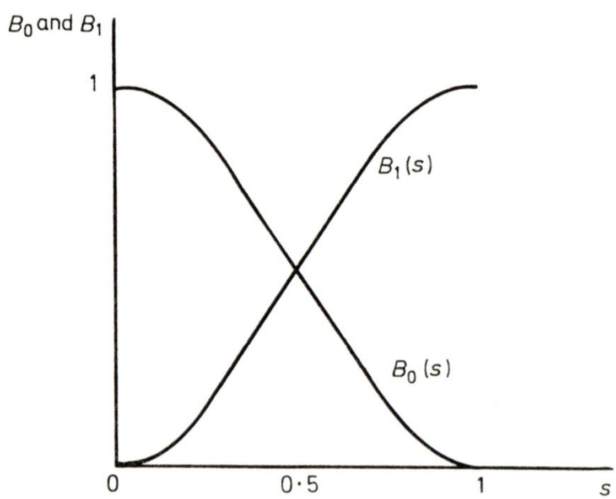

Figure 5.15 The form of the blending functions $B_0(s)$ and $B_1(s)$. The forms of $B_0(t)$ and $B_1(t)$ are identical

$$\left.\begin{aligned} \text{When } s &= 0,\ B_0(s) = 1 \text{ and } B_1(s) = 0 \\ s &= 1,\ B_0(s) = 0 \text{ and } B_1(s) = 1 \\ t &= 0,\ B_0(t) = 1 \text{ and } B_1(t) = 0 \\ t &= 1,\ B_0(t) = 0 \text{ and } B_1(t) = 1 \end{aligned}\right\} \tag{5.11}$$

These conditions are illustrated in Figure 5.15.
It should be noted that $B_1(s)$ is equivalent to $B_0(s)$ 'rotated' about $s = 0{\cdot}5$.

To verify that condition (b) is satisfied by the blending functions, consider the side $\boldsymbol{P}(0, t)$. The conditions for s and t are: $s = 0$, $0 \leq t \leq 1$, i.e.

$$B_0(s) = 1 \qquad B_0(s) = 0$$

Substituting these conditions into equation (5.9) gives

$$\begin{aligned} \boldsymbol{P}(s, t) &= \boldsymbol{P}(0, t) + \boldsymbol{P}(0, 0)\, B_0(t) + \boldsymbol{P}(0, 1)\, B_1(t) \\ &\quad - \boldsymbol{P}(0, 0)\, B_0(t) - \boldsymbol{P}(0, 1)\, B_1(t) \\ &= \boldsymbol{P}(0, t). \end{aligned}$$

Therefore, the condition that $\boldsymbol{P}(0, t)$ is contained in $\boldsymbol{P}(s, t)$ is satisfied. In the same way the remaining three sides can be shown to be contained in the surface.

An essential requirement is the joining of several patches together to form a complete surface. It follows that there must be some facility for estimating the slope at a boundary, and for matching this to an adjoining patch. When two patches have a common boundary represented by some function, it is necessary to equate the local slope of both patches in the direction normal to the boundary to maintain a smooth join.

Taking the boundaries $\boldsymbol{P}(0, t)$ and $\boldsymbol{P}(1, t)$, the slope normal to the boundary is given by

$$\frac{\partial \boldsymbol{P}}{\partial s}(s, t) \quad \text{at} \quad s = 0 \quad \text{and} \quad s = 1$$

Differentiating equation (5.9) gives

$$\begin{aligned} \frac{\partial}{\partial s} \boldsymbol{P}(s, t) &= \boldsymbol{P}(0, t)\, B_0'(s) + \boldsymbol{P}(1, t)\, B_1'(s) + \boldsymbol{P}'(s, 0)\, B_0(t) + \boldsymbol{P}'(s, 1) B_1(t) \\ &\quad - \boldsymbol{P}(0, 0)\, B_0'(s)\, B_0(t) - \boldsymbol{P}(1, 0)\, B_1'(s)\, B_0(t) \\ &\quad - \boldsymbol{P}(0, 1)\, B'_0(s)\, B_1(t) - \boldsymbol{P}(1, 1)\, B_1'(s)\, B_1(t) \end{aligned} \tag{5.12}$$

where P' and B' denote the partial derivatives with respect to s.

Consider the boundary $\boldsymbol{P}(0, t)$, i.e. $s = 0$, then from equations (5.10):

$$\text{at } s = 0, \quad B_0'(s) = 0 \quad \text{and} \quad B_1'(s) = 0$$

Substituting into (5.12) and setting $s = 0$:

$$\frac{\partial \boldsymbol{P}}{\partial s}(0, t) = \boldsymbol{P}'(0, 0)\, B_0(t) + \boldsymbol{P}'(0, 1)\, B_1(t)$$

Referring to Figure 5.12, $\boldsymbol{P}'(0, 0) = \boldsymbol{S}_1$ and $\boldsymbol{P}'(0, 1) = \boldsymbol{S}_3$

$$\frac{\partial \boldsymbol{P}}{\partial s}(0, t) = \boldsymbol{S}_1 B_0(t) + \boldsymbol{S}_3 B_1(t) \tag{5.13}$$

The same argument can be applied to the other three sides.

Equation (5.13) is an extremely important result. It shows that the slope of the surface normal to the boundary, and anywhere along that boundary, is dependent only on the blending functions and the slope vectors at either end of the boundary. Higher derivatives can, in the same way, be written in terms of the derivatives at the end points.

Summarising:

(i) An important property of Coon's patches is that continuity between patches at their corner points ensures continuity along the patch boundaries.

5.7 APPLICATION OF SURFACE PATCH THEORY

So far no mention has been made of the functional form of the bounding curves. This can be any function, but clearly the parametric form of Ferguson's Spline and Coons' Patches, when combined, produce a well matched system. The parameter U in the spline equation is directly equivalent to s and t in the surface equation, depending upon the direction in which the curve is running. It is now possible to define a complete surface as a mesh of parametric curves, the patches enclosed by the mesh being defined by the surface equation.

The way in which a surface is divided into patches dictates the accuracy of representation. In Figure 5.11 a greater concentration of patches is shown in the region where the most rapid change of shape takes place. The application of surface patch theory to real engineering problems is as much a craft as a science. For this reason the display of the surface on a CRT screen enables the designer to have an immediate visual image of the effects of any surface changes.

There is a fundamental difference in visualisation between Coons' patches and traditional lofting. Coons' patches deal with surface representations whereas lofting is concerned with projected sections. To overcome the psychological difficulties Bézier[18] developed another method which enables the form designer of a motor car to suggest alterations to the shape in a digital form. An example of the successful application of parametric splines combined with parametric surfaces is in the British Aircraft Corporation's Numerical Master Geometry system[19].

In the patchwork system described, shape is totally defined by the surface equation and the boundary curves. To modify the surface it is necessary to modify one or more of the bounding curves. An alternative method, which does not affect the boundaries, employs twist vectors. Unfortunately it is difficult for the designer to vsualise the effects of modification in this way. A fuller description of the problem is given by Armit[20].

Summarising:

(i) Coon's patches require a visualisation of surfaces rather than sectional elevations.

5.8 EXAMPLE IN SURFACE DEFINITION

As with curve description, numerical surface definition is primarily intended for manipulation on a digital computer, but the following example serves to illustrate the analysis.

EXAMPLE 3

Consider the 100 unit square in the plane $z = 0$ shown in Figure 5.16(a). The slope vectors to its edges at the corner points have only a single component of either x or y. As the edges are straight, varying their magnitudes has no effect on the form of the figure. For a start let all the slope vectors have a magnitude of 100. All corners will be arbitrarily numbered for identification and slope vectors in the general direction of x will be denoted as $\boldsymbol{S}$ and in the direction of y as $\boldsymbol{T}$; this is indicated in Figure 5.16.

If the surface at the point marked 7 ($x = 80$, $y = 60$) in Figure 5.16(b) is raised by 15 units in the z direction then from what follows it will be seen that the magnitudes of the initial slope vectors to the square have an effect on the form of the surface. The initial square, which could be considered as a single Coons' patch, is now subdivided into four patches by two lines intersecting at the raised point 7. Since $\boldsymbol{T}_1$ and $\boldsymbol{T}_2$ are identical, $\boldsymbol{T}_5$ will have the same value; this can be shown by using the appropriate form of equation (5.13), likewise for $\boldsymbol{T}_9$, $\boldsymbol{S}_6$

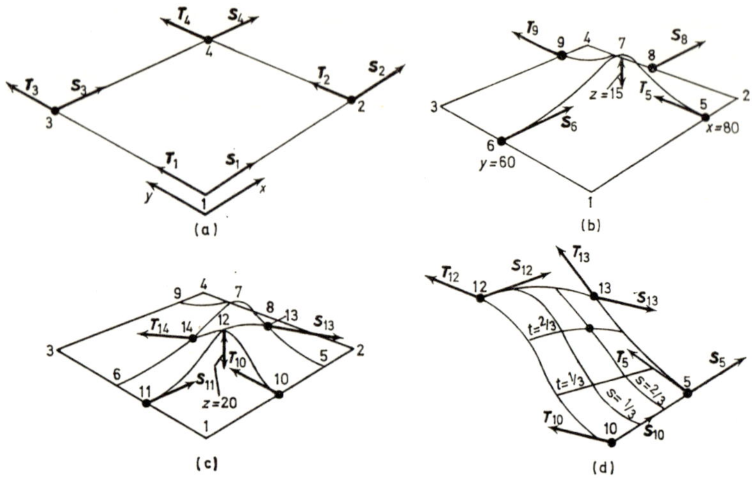

Figure 5.16 The development of a curved surface from a square plane by subdividing it into patches (vectors not to scale)

and $\boldsymbol{S}_8$. Therefore

$$\boldsymbol{T}_5 = 0\boldsymbol{i}+100\boldsymbol{j}+0\boldsymbol{k} = \boldsymbol{T}_9$$
$$\boldsymbol{S}_6 = 100\boldsymbol{i}+0\boldsymbol{j}+0\boldsymbol{k} = \boldsymbol{S}_8$$
$$\boldsymbol{P}_5 = 80\boldsymbol{i} \qquad \boldsymbol{P}_6 = 60\boldsymbol{j}$$
$$\boldsymbol{P}_7 = 80\boldsymbol{i}+60\boldsymbol{j}+15\boldsymbol{k}$$
$$\boldsymbol{P}_8 = 100\boldsymbol{i}+60\boldsymbol{j}$$
$$\boldsymbol{P}_9 = 80\boldsymbol{i}+100\boldsymbol{j}$$

The two curves intersecting at point *7* may now be treated as parametric splines having the above end-point vectors, but with their slope vectors at point 7 still to be found from equations (5.8). These equations give

$$\boldsymbol{S}_7 = 25\boldsymbol{i}+0\boldsymbol{j}+0\boldsymbol{k}$$

and

$$\boldsymbol{T}_7 = 0\boldsymbol{i}+25\boldsymbol{j}+0\boldsymbol{k}$$

Suppose the surface is raised at another position by 20 units. Take this position to be at $x = 49{\cdot}37$, $y = 39{\cdot}37$; this coincides with the point $U = 0{\cdot}5$ on the curves between points 6 and 7 and points 5 and 7 calculated from equation (5.4). Any point in x and y may have been taken, but this could have involved solving a cubic equation for U which would have been of little or no benefit to this example.

Consider the slope vector $\boldsymbol{S}_{13}$ (Figure 5.16c); this can be calculated from the values of $\boldsymbol{S}_5$ and $\boldsymbol{S}_7$. As a value of $U = 0{\cdot}5$ on the curve between 5 and 7 defines point 13, then the surface parameter t will also be 0·5 at this point. It follows from equations (5.10) that $B_0(t) = B_1(t) = 0{\cdot}5$. The appropriate form of equation (5.13) for the slope of $\boldsymbol{S}_{13}$ considering the patch 1, 6, 7, 5 is

$$\frac{\partial \boldsymbol{P}}{\partial s}(1, t) = \boldsymbol{S}_5\, B_0(t) + \boldsymbol{S}_7\, B_1(t)$$

$$\frac{\partial \boldsymbol{P}}{\partial s}(1, 0{\cdot}5) = 0{\cdot}5 \times 100\boldsymbol{i} + 0{\cdot}5 \times 25\boldsymbol{i} = 62{\cdot}5\boldsymbol{i}$$

This is the gradient only and the magnitude may be varied to affect the form of the curve passing through points 11, 12 and 13. In practice the magnitude is often set equal to a typical chord length between adjacent points, but here it will be assumed that $\boldsymbol{S}_{13} = 62{\cdot}5\boldsymbol{i}$. Similarly $\boldsymbol{T}_{14} = 62{\cdot}5\boldsymbol{j}$, $\boldsymbol{T}_{10} = 100\boldsymbol{j}$ and $\boldsymbol{S}_{11} = 100\boldsymbol{i}$, and, as previously from equations (5.8),

$$\boldsymbol{S}_{12} = 19{\cdot}37\boldsymbol{i} + 0\boldsymbol{j} + 5{\cdot}62\boldsymbol{k}$$

$$\boldsymbol{T}_{12} = 0\boldsymbol{i} + 4{\cdot}37\boldsymbol{j} + 5{\cdot}62\boldsymbol{k}$$

Every patch in Figure 5.16(c) can now be defined by slope vectors to the curve at its corner points, some of which are shared by adjoining patches and at the same time the patches are joined smoothly together; this was established by equation (5.13). Furthermore each patch has a unique surface equation from which the coordinates of any point on the surface can be calculated. This will be illustrated by considering the patch in Figure 5.16(d). The vectors for the patch corners are

$$\boldsymbol{P}_{10} = 49{\cdot}37\boldsymbol{i} \quad \boldsymbol{P}_5 = 80\boldsymbol{i} \qquad \boldsymbol{P}_{12} = 49{\cdot}37\boldsymbol{i} + 39{\cdot}37\boldsymbol{j} + 20\boldsymbol{k}$$

$$\boldsymbol{P}_{13} = 80\boldsymbol{i} + 39{\cdot}37\boldsymbol{j} + 7{\cdot}5\boldsymbol{k}$$

Taking the origin ($s = t = 0$) at point 10, the coordinates of the surface at the position $s = t = \frac{2}{3}$ will be calculated. The curve between points 5 and 7 has already been established; the position $t = \frac{2}{3}$ along the side between points 5 and 13 corresponds to the position $U = \frac{2}{6} = \frac{1}{3}$ on the curve between points 5 and 7 giving a coordinate vector: $(80\boldsymbol{i} + 18{\cdot}5\boldsymbol{j} + 3{\cdot}889\boldsymbol{k})$. This corresponds to the side $P(1, t)$ in equation (5.9).

Therefore $\qquad \boldsymbol{P}(1, \frac{2}{3}) = 80\boldsymbol{i} + 28{\cdot}5\boldsymbol{j} + 3{\cdot}889\boldsymbol{k}$

similarly $\qquad \boldsymbol{P}(0, \frac{2}{3}) = 49{\cdot}37\boldsymbol{i} + 35{\cdot}92\boldsymbol{j} + 13{\cdot}98\boldsymbol{k}$

$$\boldsymbol{P}(\tfrac{2}{3}, 0) = 80\boldsymbol{i}$$

$$\boldsymbol{P}(\tfrac{2}{3}, 1) = 64{\cdot}23\boldsymbol{i} + 39{\cdot}37\boldsymbol{j} + 11{\cdot}15\boldsymbol{k}$$

From equations (5.10), when $s = t = \frac{2}{3}$

$$B_0(s) = B_0(t) = 0{\cdot}259$$

and

$$B_1(s) = B_1(t) = 0{\cdot}741$$

Substituting into equation (5.9):

$$\begin{aligned}\boldsymbol{P}(\tfrac{2}{3}, \tfrac{2}{3}) = {} & 0{\cdot}259(49{\cdot}37\boldsymbol{i}+35{\cdot}92\boldsymbol{j}+13{\cdot}98\boldsymbol{k})+0{\cdot}741(80\boldsymbol{i}+28{\cdot}5\boldsymbol{j}+3{\cdot}88\boldsymbol{k}) \\ & +0{\cdot}259\times 80\boldsymbol{i}+0{\cdot}741(64{\cdot}23\boldsymbol{i}+39{\cdot}37\boldsymbol{j}+11{\cdot}15\boldsymbol{k}) \\ & -0{\cdot}259\times 0{\cdot}259\times 49{\cdot}37\boldsymbol{i}-0{\cdot}741\times 0{\cdot}259\times 80\boldsymbol{i} \\ & -0{\cdot}259\times 0{\cdot}741(49{\cdot}37\boldsymbol{i}+39{\cdot}37\boldsymbol{j}+20\boldsymbol{k}) \\ & -0{\cdot}741\times 0{\cdot}741(80\boldsymbol{i}+39{\cdot}37\boldsymbol{j}+7{\cdot}5\boldsymbol{k})\end{aligned}$$

Multiplying out and adding like components yields

$$\boldsymbol{P}(\tfrac{2}{3}, \tfrac{2}{3}) = 67\boldsymbol{i}+31{\cdot}46\boldsymbol{j}+7{\cdot}09\boldsymbol{k}$$

By calculating further values, lines of constant s and t can be drawn as shown in Figure 5.16(d). Although the calculations are simple it can be seen that they are also numerous and tedious thus enforcing the argument for the use of the digital computer. A further problem arises from this example and that is the complexity of redefining origins as a consequence of subdividing patches. To program this method for use on a digital computer requires a high standard of data organisation.

It will have been noticed that the lines drawn on the patch in Figure 5.16(d) are lines of constant s and t and not lines of constant x or y. The patch method is very much oriented towards surface location rather than the x, y coordinate frame of reference. If a value of z is required for particular values of x and y it is first necessary to find the corresponding value of U from equation (5.4); this leads to a cubic equation for U. Having found U, the method proceeds as in this example. This is the exercise which must be carried out in order to take vertical or horizontal contour sections through the surface.

6 Some Aspects of Optimisation

6.0 WHAT IS OPTIMISATION?

Optimisation is the search for the solution from which the 'maximum benefit' will be gained. The designer is always seeking an optimum solution, this is the aim of design methodology, but the discussion in this chapter will be limited to optimising relationships between various parameters, which may or may not be related mathematically.

The expression 'maximum benefit' is carefully chosen. In some situations, e.g. the power available from an engine, the designer may be looking for the maximum value, whereas with product cost he may seek a minimum. Mischke[21] defines a merit function linking the various parameters. The function is so arranged that when it has a maximum value the optimum situation exists. This means that the parameters for which a minimum is desirable have to be arranged within the function so as to cause an increase in the function with a decrease in the parameters. In most design situations it is sufficient to maximise or minimise the functional relationship.

It will be assumed initially that a mathematical relationship exists between the parameters in the system, this helps to simplify the understanding of the problem, but many of the arguments are still valid although the relationship may be unknown.

Consider the function $f(x)$ plotted against x in Figure 6.1(a). In the shaded region bounded by x_{min} and x_{max} the function has a well defined maximum value at A; if a minimum value should be the optimum this would occur at B. It may seem artificial to limit x in this way, but this is a real situation; in practice the designer is limited by the maximum or minimum sizes of raw material or by the output usefully consumed from a power station, etc. This shaded area is termed the *feasible region*. Sometimes optimisation becomes meaningless if a feasible region is not specified. Figure 6.1(b) shows one such example; the maximum in $f(x)$ for the unbounded curve is infinite. The maximum

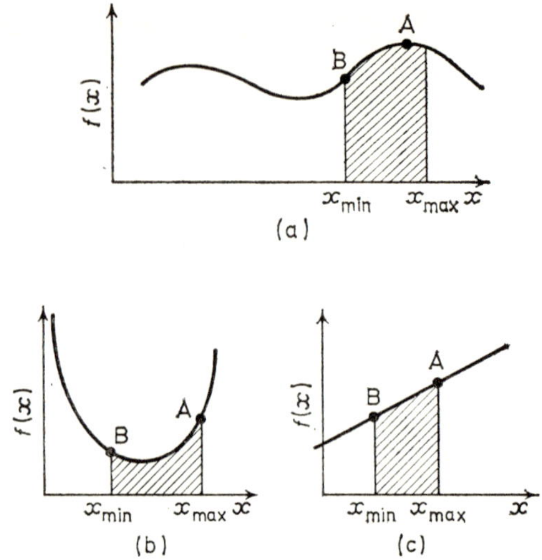

Figure 6.1 The effect of the feasible region on maxima and minima

for the feasible region is at A. If the lower boundary is moved to the left, eventually B becomes the maximum. Figure 6.1(c) shows a linear function; a maximum exists at A and a minimum at B. These must always occur at the boundaries of the feasible region, hence an optimum solution can never be found for an unbounded linear function. This is the basis of linear programming.

Summarising:

(i) All design variables have upper and lower limits.

6.1 DIFFICULTIES WITH CLASSICAL OPTIMISATION

A system may be described by a mathematical relationship, and for optimisation it can be arranged so that the variable to be maximised or minimised becomes a function of the other variables. These variables may be independent or related in some way. For independent variables, one way of optimising is to take partial derivatives with respect to each variable and equate to zero. This gives a set of simultaneous equations of the form

$$\frac{\partial f}{\partial x_1} = \frac{\partial f}{\partial x_2} = \ldots = \frac{\partial f}{\partial x_i} = \ldots \frac{\partial f}{\partial x_N} = 0 \quad \text{where} \quad f = f(x_1, x_2 \ldots x_N)$$

Whether f is a maximum or a minimum depends on the sign of the second derivative.

The shortcomings of this method are:

(i) Continuous first and second derivatives of the function must exist.

(ii) This set of equations must be solved, which could be no mean task; one method is to minimise the function

$$\psi = \sum_{i=1}^{N} \left(\frac{\partial f}{\partial x_i}\right)^2$$

which is selfdefeating.
(The power takes account of positive and negative values of the derivatives.)

(iii) The equation can be satisfied by a mixture of maxima and minima of the function with respect to its variables, as shown in Figure 6.2 (i.e. saddle points).

If the variables are related, in theory, it would appear that the problem could be reduced to one of finding the maximum or minimum of a function with one independent variable. In practice, this is not always possible, and the Lagrangian method of undetermined multipliers is used.

This method can be extended to any number of variables, but the complexity involved makes it unattractive.

This does not mean that the classical methods have no use at all. For simple problems involving a few variables and a well defined continuous functional relationship they may prove to be the most satisfactory. This raises a general question: Which is the best method of optimisation? There is no method which can be universally applied.

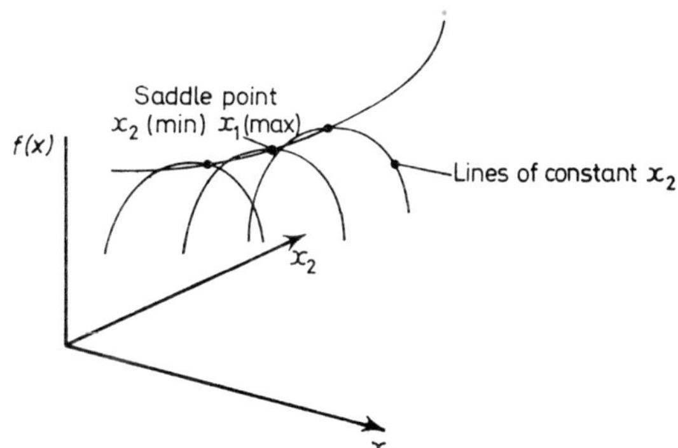

Figure 6.2 A saddle point on a function of two variables

The form of the problem may suggest a particular approach; such is the case with linear relationships, but often optimisation becomes a 'trial-and-error' science.

Summarising:

(i) There is no one method of optimisation which can be universally applied.

6.2 LINEAR PROGRAMMING

The problem of optimising linear relationships was first posed in the 1930s; this was apparent from trivial one-dimensional and two-dimensional examples such as the one in Figure 6.1(c). More general solutions of that time were limited to specific aspects such as the transportation problem and its solution by the stepping-stone method, to be described later. It was not until the mid 1940s, when Dantzig[22] introduced the Simplex Method, that universal problems, involving several variables, could be tackled.

The Simplex Method is a particularly powerful form of optimisation. It can handle any number of variables and will always converge on the maximum or minimum value for the specified region; the only constraints are the time and patience of the user. The method is particularly attractive for use on the computer since the mathematical manipulations are simple, though plentiful. Minor difficulties which could cause instability are rectified by checks built into the program. Nearly all computer manufacturers provide program packages for linear programming; it is only necessary for the user to provide the data in a specific order.

Although linear programming is often performed on the computer, the Simplex Method can be used in the form of a tableau for hand working. This is particularly attractive to the designer handling problems with few variables. The form of tableau varies extensively. Some of the tableaux become quite complex but provide a check on errors and an indication of progress. The purpose of this text is to show how engineering problems can be organised into a programmable form. For this reason a particularly basic form of the tableau has been adopted for simplicity.

From Figure 6.1(c), the description of the problem falls into two categories; the linear function passing through A and B, and the defining of the feasible region. Figure 6.1(c) is one-dimensional showing a linear function comprising one independent and one dependent variable. In practice the number of independent variables is unlimited. For complex examples this is often around the thousand mark. The function to be optimised is known as the *objective* function and in the general case with n independent variables takes the form

$$F = \sum_{j=1}^{n} c_j x_j \qquad (j = 1, 2 \ldots n)$$

A further requirement for linear programming is

$$x_j \geq 0$$

A way of getting around this constraint will be described later.

The feasible region is described by a set of inequalities illustrated by the following example. The example is two dimensional, having independent variables x_1 and x_2 and a dependent variable F:

$$-x_1+3x_2 \leq 10 \qquad (1)$$

$$x_1+\ x_2 \leq 6 \qquad (2)$$

$$x_1-\ x_2 \leq 2 \qquad (3)$$

$$x_1+3x_2 \geq 6 \qquad (4)$$

$$x_1+2x_2 = F = \max \qquad (5)$$

Equations (1) to (4) give the region, and equation (5) is the objective function.

Had the minimum value of the objective function been required, then (5) becomes

$$-x_1-2x_2 = \max$$

The inequalities in equations (1) to (4) can be made equalities by introducing *slack* variables x_3, x_4, x_5 and x_6, giving

$$-x_1+3x_2+x_3 = 10 \qquad (1a)$$

$$x_1+\ x_2 \quad +x_4 = 6 \qquad (2a)$$

$$x_1-\ x_2 \quad +x_5 = 2 \qquad (3a)$$

$$x_1+3x_2 \quad -x_6 = 6 \qquad (4a)$$

The slack variables give an indication of where, relative to the limiting boundary, a point is placed within the feasible region, e.g. when the slack variable is zero, the boundary has been reached, when it is negative the boundary has been violated. Equations (1) to (4) define the region shown in Figure 6.3.

The line $x_3 = 0$ is AB, i.e. $-x_1+3x_2 = 10$, $x_4 = 0$ is BC, $x_5 = 0$ is CD, $x_6 = 0$ is DE and $x_1 = 0$ is EA.

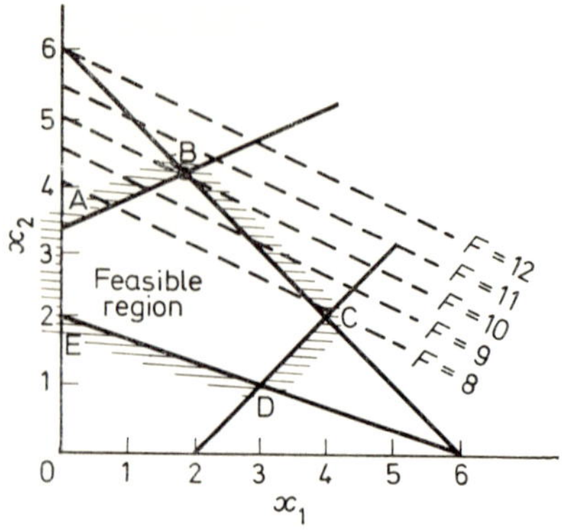

Figure 6.3 The graphical representation of equations (1) to (5)

Values of the objective function F can be plotted as a series of parallel lines on Figure 6.3. This is done by assuming a value for F, taking two values for x_1, and calculating the two corresponding values of x_2, giving two points through which that line must pass. From the diagram the maximum value of F within the feasible region is $F = 10$, giving $x_1 = 2$ and $x_2 = 4$ at the vertex B, i.e. the line representing the function F has been moved across the graph to the point where it leaves the feasible region.

This leads to the very important observation: the optimum value of a linear function always occurs at a vertex of the feasible region. Had the line representing the function F been parallel to BC then an optimum would have existed at the two vertices B and C, and anywhere along BC; in which case either B or C would be acceptable,

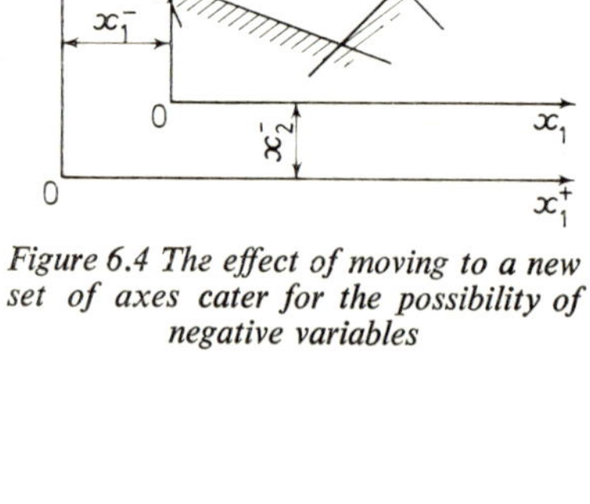

Figure 6.4 The effect of moving to a new set of axes cater for the possibility of negative variables

though in practice a preference for one or the other might be indicated.

Consider the situation where x_1 and x_2 are not always positive. The restriction can be overcome by setting

$$x_1 = x_1^+ - x_1^- \text{ or } x_2 = x_2^+ - x_2^-$$

where x_1^- can be greater than x_1^+. This is equivalent to moving from the original set of axes to a new frame of reference, x_1^+ and x_2^+, as illustrated in Figure (6.4) for the possibility of the previous example having negative variables.

Summarising:

(i) A linear function must be bounded for an optimum to exist.
(ii) Optima always occur at a vertex to the feasible region.
(iii) The Simplex method will converge regardless of the number of variables.

6.2.1 The Simplex Tableau

The graphical method already described for optimising a linear function is only realistic for two-dimensional problems; the observation that an optimum always occurs at a vertex applies to any number of dimensions, although it is difficult to visualise a vertex in more than three dimensions. To deal with multidimensional problems the Simplex Method and tableau devised by Dantzig are employed.

The method will be illustrated by the following three-dimensional example:

$$x_1 + 2x_2 + 2x_3 = \max \tag{1}$$

subject to

$$2x_1 + x_2 + 3x_3 \leq 6 \tag{2}$$

$$4x_1 + 2x_2 + 3x_3 \leq 8 \tag{3}$$

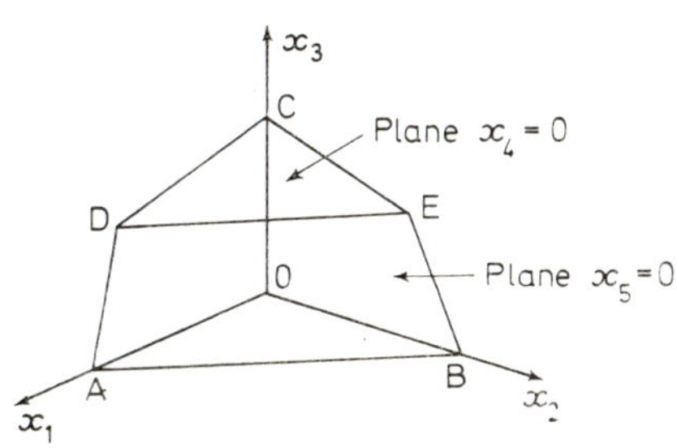

Figure 6.5 The bounding planes to a three-dimensional feasible region

Equations (2) and (3) define the solid region shown in Figure 6.5; the bounding lines to the feasible region now become bounding planes. Introducing the slack variables:

$$2x_1 + x_2 + 3x_3 + x_4 \qquad = 6 \qquad (2a)$$

$$4x_1 + 2x_2 + 3x_3 \qquad + x_5 = 8 \qquad (3a)$$

There are now five bounding planes: those of the axes $x_1 = 0$, $x_2 = 0$, $x_3 = 0$ and the two planes defined by $x_4 = 0$ (CDE) and $x_5 = 0$ (ABED).

Rearranging (2a) and (3a):

$$x_4 = 6 - 2x_1 - x_2 - 3x_3 \qquad (2a)$$

$$x_5 = 8 - 4x_1 - 2x_2 - 3x_3 \qquad (3a)$$

and the objective $= x_1 + 2x_2 + 2x_3$ (1)

Since a vertex is being sought it would seem reasonable to examine each vertex in turn until an optimum is found. Starting at the vertex $x_1 = x_2 = x_3 = 0$, i.e. the origin, it is evident that the objective function is not a maximum. If x_3 is increased the search moves along the line OC to the vertex C where $x_1 = x_2 = x_4 = 0$, but to recalculate the objective function it is necessary to write it in terms of x_4 instead of x_3 which is now unknown. To simplify the rewriting of the equations for each vertex, they are represented in a tabular form. At any vertex some of the variables will be zero, these are known as *nonbasic variables*, the remainder being *basic variables*.

The tableau for the vertex at the origin, representing equations (1), (2a) and (3a), is written as:

Basic variables	*b*	*Nonbasic variables* x_1	x_2	x_3
x_4	6	2	1	3
x_5	8	4	2	3
obj	0	−1	−2	−2

Note the change in signs. The signs in the tableau are equivalent to:

$$2x_1 + x_2 + 3x_3 + x_4 = 6$$

$$4x_1 + 2x_2 + 3x_3 + x_5 = 8$$

$$-x_1 - 2x_2 - 2x_3 + \text{obj} = 0$$

When the move is made to vertex C, equations (1), (2a) and (3a) are rearranged to have the objective function in terms of x_1, x_2 and x_4, giving

$$x_3 = 2 - 2/3x_1 - 1/3x_2 - 1/3x_4$$
$$x_5 = 2 - 2x_1 - x_2 + x_4$$
$$\text{obj} = 4 - 1/3x_1 + 4/3x_2 - 2/3x_4$$

The new tableau about C becomes:

Basic	b	x_1	x_2	x_4
x_3	2	$\frac{2}{3}$	$\frac{1}{3}$	$\frac{1}{3}$
x_5	2	2	1	−1
obj	4	$\frac{}{3}$	$-\frac{4}{3}$	$\frac{2}{3}$

It can now be seen that the objective function has increased in value, but by the method described it would be necessary to test each vertex in turn until a maximum is found. The Simplex method operates on the tableau, but guides the user along the edge which will give him an improvement in the objective function; in this example it was not evident that OC was the best edge to move along—OB or OA may have been better. The method also shows the user when the optimum has been reached. The Simplex method is a mechanical way of handling the tableau as follows: (taking a particular element in the main part of the tableau as a_{ij} being on row i and in column j and b_i as an element in the b column)

1 Set up a Simplex tableau about zero, letting the slack variables be the basic variables.
2 Find the largest negative nonbasic element in the objective row. This is the best nonbasic variable to increase.
3 If this variable exists in column q, then find the row in which b_i/a_{iq} has the least value and a_{iq} is positive. The basic variable in this row (row p) is made a nonbasic (zero) element in the next tableau.
   Note: The edge along which the search is going to move, and the plane that this edge intersects to form the next Simplex tableau, is defined by step 2.
4 The intersection of row p with column q defines the pivot for the next tableau, the rearrangement of the equations can now be expressed in a mechanical form.
   Let the corresponding element to a_{ij} on the new tableau be a'_{ij}. (Note that a and a' are now taken to be general and include

the values in column b.) Then

$$a'_{pq} := \frac{1}{a_{pq}}$$

$$a'_{iq} := \frac{-a_{iq}}{a_{pq}} \qquad (i \neq p)$$

$$a'_{pj} := \frac{a_{pj}}{a_{pq}} \qquad (j \neq q)$$

and

$$a'_{ij} := a_{ij} - \frac{a_{iq}a_{pj}}{a_{pq}} \qquad (i \neq p, j \neq q)$$

5 Operations from 2 onwards are repeated until no negative element in the objective row remains; when this is so, the optimum has been reached.

The Simplex tableaux representing this problem (where the circle indicates the pivot point for the next tableau) are:

	b	x_1	x_2	x_3	
x_4	6	2	1	$\textcircled{3}$	Tableau about 0
x_5	8	4	2	3	$x_1 = x_2 = x_3 = 0$
obj	0	-1	-2	-2	

	b	x_1	x_2	x_4	
x_3	2	$\frac{2}{3}$	$\frac{1}{3}$	$\frac{1}{3}$	
x_5	2	2	$\textcircled{1}$	-1	Tableau about C
obj	4	$\frac{1}{3}$	$-\frac{4}{3}$	$\frac{2}{3}$	

	b	x_1	x_5	x_4	
x_3	$\frac{4}{3}$	0	$-\frac{1}{3}$	$\textcircled{\frac{2}{3}}$	
x_2	2	2	1	-1	Tableau about E
obj	$6^2/3$	3	$\frac{4}{3}$	$-\frac{2}{3}$	

	b	x_1	x_5	x_3	
x_4	2	0	$-\frac{1}{2}$	$\frac{3}{2}$	
x_2	4	2	$\frac{1}{2}$	$\frac{3}{2}$	Tableau about B
obj	8	3	1	1	

Note that the two variables earmarked by the pivot point change places in the following tableau.

The final nonbasic variables are x_1, x_5 and x_3. Therefore $x_1 = x_5 = x_3 = 0$ which gives, from (3a), $x_2 = 4$, from (1) the maximum value of the objective function is 8, and from (2a) the value of the slack variable $x_4 = 2$. Graphically, the optimum exists at point B, and to reach it the search has gone from 0 to C, C to E and E to B.

6.2.2 Starting the Search

In the last example it was seen from Figure 6.5 that the origin of the three coordinates was a vertex to the feasible region. If this origin had been excluded from the region by a further bounding plane the search must then start from another vertex. With two-dimensional and three-dimensional problems a suitable vertex can be found by inspection; for multidimensional analyses a formalised method must be adopted, this is known as two-phase searching. Phase one is to find any vertex to the feasible region whilst phase two is aimed at finding the optimum. Computer software is normally written to automatically carry out a two-phase search regardless of the problem; the more sophisticated Simplex Tableaux cater for the user in the same way. As this text is primarily concerned with introducing the reader to the application of linear programming to engineering problems, phase one will be limited to a brief discussion of the principles involved.

Consider the first example illustrated in Figure 6.3; the value $x_1 = x_2 = 0$ falls outside the feasible region.

Four new variables are introduced into the equations:

$$\begin{aligned}
-x_1+3x_2+x_3 \quad &+U_1 &&= 10\\
x_1+x_2 \quad +x_4 \quad &+U_2 &&= 6\\
x_1+x_2 \quad +x_5 \quad &+U_3 &&= 2\\
x_1+3x_2 \quad -x_6 \quad &+U_4 &&= 6
\end{aligned}$$

These are known as *artificial variables.*

The objective function is replaced by a new objective function of the artificial variables:

$$U_1+U_2+U_3+U_4 = z$$

The exercise becomes one of minimising z, and the initial tableau may be based on

$$x_1 = x_2 = \ldots x_6 = 0$$

i.e.

$$\begin{aligned}
U_1 &= 10\\
U_2 &= 6\\
U_3 &= 2\\
U_4 &= 6
\end{aligned}$$

If the minimum value of $z \neq 0$ then a feasible solution does not exist.

Alternatively the objective function may be retained in its original form with the addition of the artificial variables:

$$-x-2x_2+MU_1+MU_2+MU_3+MU_4 = \min$$

where M is very large. The effect of M being large is to drive the values of U to zero, having the same effect as minimising z in the previous method.

6.2.3 Some Observations on Linear programming

Working by hand methods

(a) It is better to find the new objective row first. If this is non-negative, the search can be terminated immediately.
(b) To simplify calculations a common factor may be removed from all the elements of the tableau.

Degeneracy

This may occur when searching for a pivotal point in a tableau. Having fixed the column in which the pivotal point will exist, it is then necessary to locate the row by minimising the ratio mentioned earlier. If there are two or more possible rows with the same value for the ratio, then one might select between them at will. If this is the case then the row which was not selected will contain a basic element which will be driven to zero at the same time as the selected basic element, meaning that there is a zero element in the basic column; this is unsatisfactory.

Degeneracy can drive the iterative process into a loop, but most linear programming computer software has a built-in test for this occurring, and generally convergence can be assured.

6.2.4 Arranging a problem for linear programming

Some nonlinear examples are easily converted into exact linear relationships.

Consider the function f given by

$$f = a_1^{c_1} \times a_2^{c_2} \times \ldots \times a_n^{c_n}$$

then

$$\log f = c_1 \log a_1 + c_2 \log a_2 + \ldots c_n \log a_n$$

Let

$$\log f = y, \quad \log a_1 = x_1, \quad \log a_2 = x_2, \quad \text{etc.}$$

then

$$y = c_1x_1 + c_2x_2 + c_3x_3 + \ldots c_nx_n$$

which is a linear form.

It must be remembered that constraints originally expressed in terms of a must be transformed into the new variables x before being written as linear inequalities.

Although some functional relationships are not linear and cannot be arranged into exact linear forms, it is sometimes worthwhile making

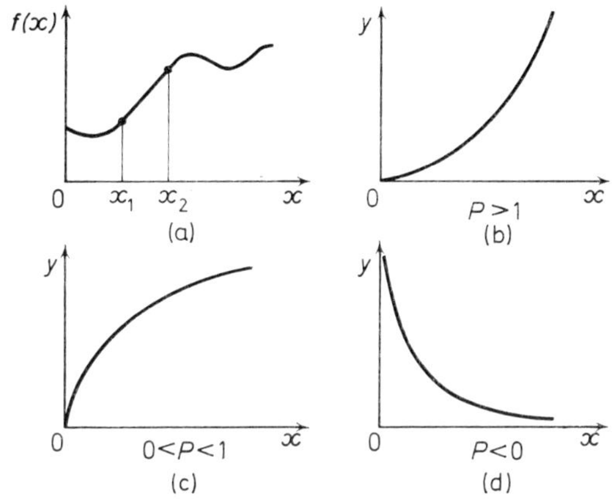

Figure 6.6 Some approximations which may be made to linearise an objective function

linear approximations. Figure 6.6(a) shows a function which is approximately linear between x_1 and x_2, i.e. in the feasible region.

A useful method of linearly approximating a function to a single variable, when the normal relationship is nonlinear, is to approximate the curve to an exponential of the form

$$y = cx^d$$

Ignoring the trivial case of $p = 1$, i.e. a linear relationship, there are three possible forms to the curve; these are shown in Figures 6.6(b) to (d).

Having made the approximation, it is then possible to write it as

$$\log y = p \log x + \log c$$

If y is a function of variables other than x, with a relationship not in the form of products or exponentials, then nothing has been gained.

Sometimes a linear function is not continuous, e.g. a tensile load can be linearly related to a simple stress and the area, but the material is only available in discrete sizes. This problem may be handled by a form of linear programming known as integer programming. Examples of this are given in reference 23.

Summarising:

(i) If a function can be linearised the optimisation is simplified considerably.

6.2.5 Examples and Questions in Linear programming

EXAMPLE 1

Use the Simplex tableau to evaluate the problem shown graphically in Figure 6.3. As the intersection of the two axes is outside the feasible region, start the search at vertex E.

Solution

Introducing slack variables:

$$-x_1+3x_2+x_3 \qquad\qquad = 10 \qquad (1)$$

$$x_1+\ x_2 \qquad +x_4 \qquad = \ 6 \qquad (2)$$

$$x_1-\ x_2 \qquad\qquad +x_5 = \ 2 \qquad (3)$$

$$x_1+3x_2 \qquad\qquad -x_6 = \ 6 \qquad (4)$$

Nonbasic variables defined by E are x_1 and x_6.

Then from (4)

$$x_2 = 2-\tfrac{1}{3}x_1+\tfrac{1}{3}x_6$$

from (3)

$$x_5 = 4-\tfrac{4}{3}x_1+\tfrac{1}{3}x_6$$

from (2)

$$x_4 = 4-\tfrac{2}{3}x_1-\tfrac{1}{3}x_6$$

from (1)

$$x_3 = 4+2x_1-x_6$$

and

$$\text{obj} = 4+\tfrac{1}{3}x_1+\tfrac{2}{3}x_6$$

The initial tableau about E is:

(E)

basic	b	x_1	x_6
x_2	2	$\frac{1}{3}$	$-\frac{1}{3}$
x_3	4	-2	①
x_4	4	$\frac{2}{3}$	$\frac{1}{3}$
x_5	4	$\frac{4}{3}$	$-\frac{1}{3}$
obj	4	$-\frac{1}{3}$	$-\frac{2}{3}$

About (A)

basic	b	x_1	x_3
x_2	$\frac{10}{3}$	$-\frac{1}{3}$	$\frac{1}{3}$
x_6	4	-2	1
x_4	$\frac{8}{3}$	$\frac{4}{3}$	$-\frac{1}{3}$
x_5	$\frac{16}{3}$	$\frac{2}{3}$	$\frac{1}{3}$
obj	$\frac{20}{3}$	$-\frac{5}{3}$	$\frac{2}{3}$

(B)

basic	b	x_4	x_3
x_2	12	$\frac{1}{4}$	$\frac{3}{4}$
x_6	8	$\frac{1}{2}$	$\frac{5}{2}$
x_1	2	$\frac{1}{4}$	$-\frac{1}{4}$
x_5	12	$-\frac{1}{2}$	$\frac{3}{2}$
obj	30	$\frac{5}{4}$	$\frac{3}{4}$

EXAMPLE 2

Shown in Figure 6.7 is an elastic system for operating both an electrical alarm contact and a mechanical indicator as a consequence of load W. The contact operates from the movement at A, and the indicator from the deflection at B. The maximum movement capable of being recor-

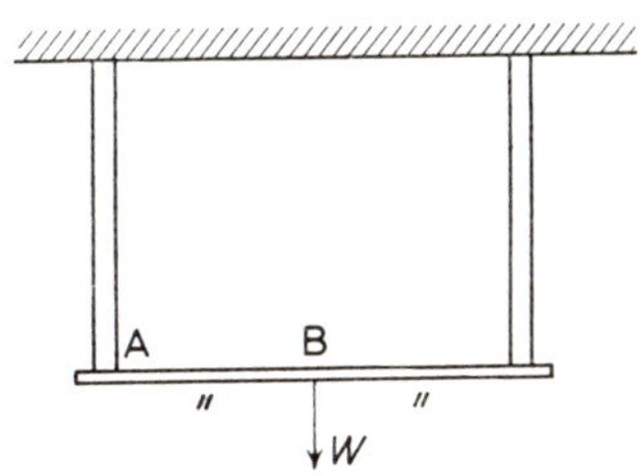

Figure 6.7 The arrangement for Example 2

ded at B is 2 mm, and the maximum permitted movement at A is 0·12 mm. The mechanical recorder requires at least 11 times the movement of the electrical contact. What is the maximum displacement which can be recorded at B within these limits?

Solution

Let deflection at A due to the pillars be x_1
Let deflection due to beam be x_2
Total deflection at B $= x_1+x_2$

The first constraint on the movement at B is $x_1 + x_2 \leq 2$, and at A $x_1 \leq 0{\cdot}12$

From the ratio of movements $\quad x_2 \geq 10x_1$

therefore $\quad x_2 - 10x_1 \geq 0$

The problem is

$$\begin{aligned} x_1 + x_2 &\leq 2 \\ x_1 \quad &\leq 0{\cdot}12 \\ -10x_1 + x_2 &\geq 0 \\ x_1 + x_2 &= \max \end{aligned}$$

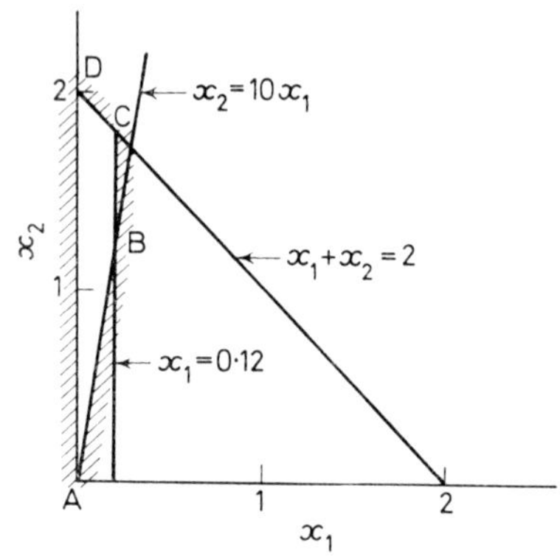

Figure 6.8 The graphical representation of Example 2

The graphical solution is shown in Figure 6.8. The solution may lie anywhere on the line joining the two vertices C and D. Clearly the vertex D, i.e. $x_1 = 0$, would be useless for operating the contact. On the other hand if a large movement was required for the contact vertex C would produce the optimum solution.

EXAMPLE 3

A fluid motor when tested in the factory has a maximum performance indicated on the torque–speed diagram in Figure 6.9 by the line A. In service the maximum performance, limited by the supply pressure, is given by line B. There is also an upper torque limit of 1200 N m and a speed limit of 1000 rev/min. Line A can be represented by the function

$$T = 2120N^{-2/3}$$

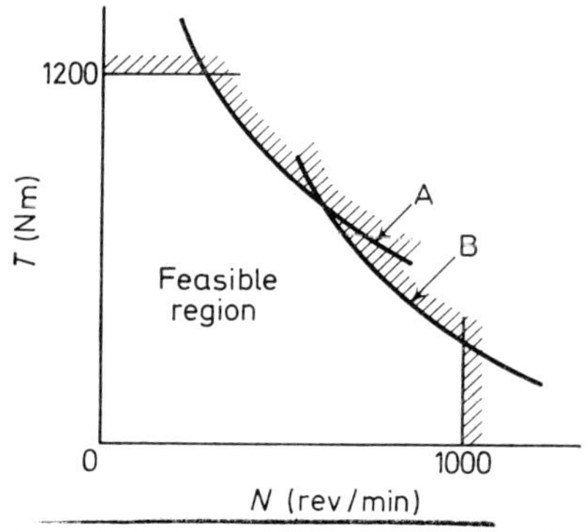

Figure 6.9 The torque–speed constraints for Example 3

and line B by

$$T = 10\times 10^4 N^{-3/2}$$

where T is in N m and N is in rev/min. Draw a graph of log N against log T and find graphically the speed of operation which produces the maximum power under service conditions. Set up the Simplex tableau about the point log T = log N = 0, leaving the constants in logarithmic form.

Solution

Log–log 5×3 cycles graph paper is required (this enables line B to cross the vertical axis to make plotting easier).

Objective function

$$\text{Power } (P) = \frac{2\pi NT}{60}$$

$$\log P = \log\frac{2\pi}{60} + \log N + \log T = \text{obj}$$

$$\text{obj} = \log\frac{2\pi}{60} + x_1 + x_2$$

Constraints

$$\tfrac{3}{2}x_1 + x_2 \leq \log 100\,000 \quad \text{(line A).}$$

$$\tfrac{2}{3}x_1 + x_2 \leq \log 2120 \quad \text{(line B)}$$

$$x_1 \leq \log 1000 \quad \text{(speed limit)}$$

$$x_2 \leq \log 1200 \quad \text{(torque limit)}$$

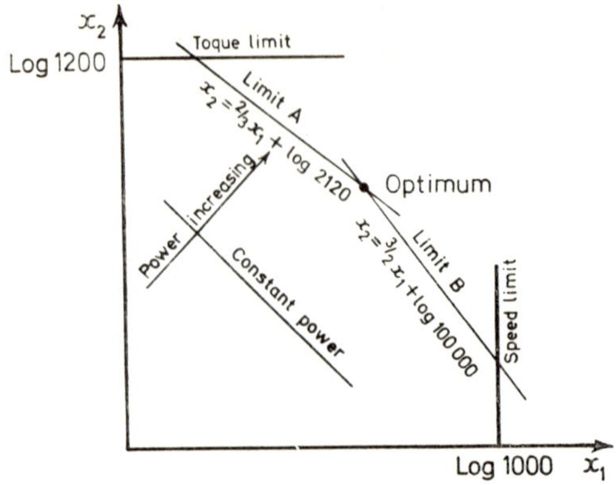

Figure 6.10 The representation of Figure 6.9 in a logarithmic form

These constraints together with the objective function are shown in Figure 6.10.

If the line of constant power is moved across the graph, from left to right, it leaves the feasible region at the point marked optimum,

$$N_{\text{OPT}} = 100 \text{ rev/min}$$

Simplex tableau

Introducing slack variables:

$$\begin{aligned} \tfrac{3}{2}x_1+x_2+x_3 \quad &= \log 100\,000 \\ \tfrac{2}{3}x_1+x_2 \quad +x_4 \quad &= \log 2120 \\ x_1 \quad +x_5 \quad &= \log 1000 \\ x_2 \quad +x_6 &= \log 1200 \end{aligned}$$

$$-x_1-x_2+\text{obj} = \log\frac{2\pi}{60}$$

The tableau about $x_1 = x_2 = 0$ is:

Basic	*Constant*	x_1	x_2
x_3	log 1 000 000	$\frac{3}{2}$	1
x_4	log 2120	$\frac{3}{2}$	1
x_5	log 1000	1	0
x_6	log 1200	0	1
obj	$\log\left(\frac{2\pi}{60}\right)$	-1	-1

EXAMPLE 4

A power boat steering system comprises a rudder shaft to the top of which is attached a 250 mm diameter drum A (Figure 6.11) A cable is wrapped around the drum and the two ends are attached to the two bell-crank levers marked B and C. A second cable is wrapped around a 100 mm diameter drum D and its ends are attached

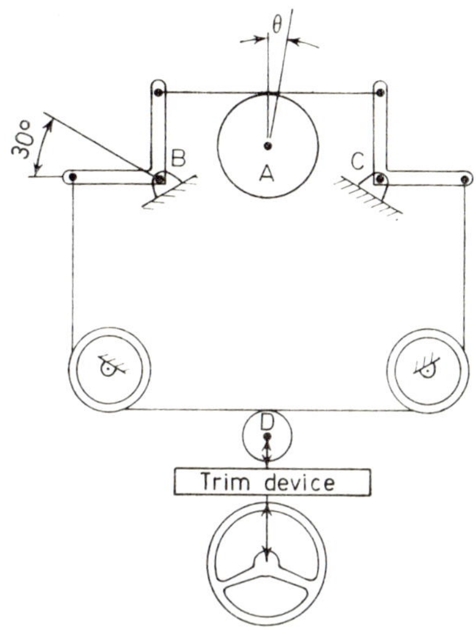

Figure 6.11 Diagrammatic layout for Example 4 (not to scale)

to the bell-crank levers as shown. This drum is connected through a trimming device to the steering wheel. The gear ratio between the drum and steering wheel is 1 : 1.

The maximum permissible movement of the steering wheel is 160° to either side of the central position. The trimming device allows a maximum course correction to be made to the rudder equivalent to $\pm 12°$ of steering wheel movement at any setting. Each bell-crank arm is 300 mm in length and their maximum allowable movement (consistent with cable tensions) is $\pm 30°$. The play in the steering wheel permits movement through a total angle of 8° without moving the rudder.

Write these constraints as mathematical inequalities, together with the objective function. Show graphically the limitation on rudder movement, and hence calculate its maximum movement.

Solution

Let the half-angle movement of the rudder be θ, the movement of the steering wheel be θ_1, and trimmer θ_2.
Objective function:

$$\theta = \tfrac{100}{250}\,(\theta_1+\theta_2-4)$$

Note: θ_2 can be $\pm$ but since maximum is sought, $+$ is taken.

Constraints on steering wheel: $\theta_1 \leq 160°$

Constraints on trimmer: $\theta_2 \leq 12°$

Constraints on bell cranks: $\dfrac{100}{2\times 300}\,(\theta_1+\theta_2-4) \leq 30°$

From the graph the feasible region is bounded by the limitations on θ_1 and θ_2, i.e. $\theta_1 = 160°$, $\theta_2 = 12°$
hence $\theta = \frac{100}{250}\,(160+12-4) = 70{\cdot}4°$
Maximum movement $= 70{\cdot}4°$ either side of centre.

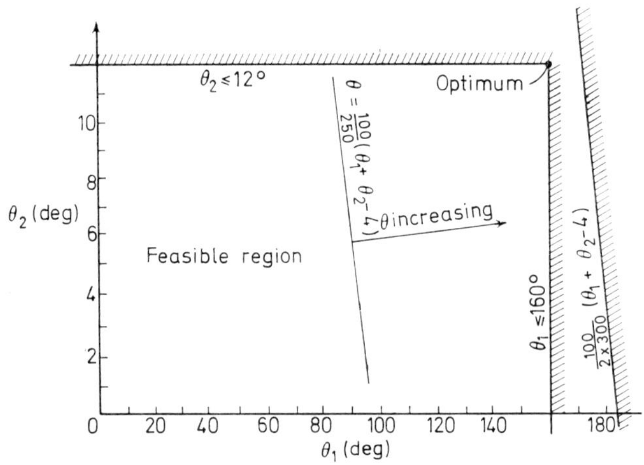

Figure 6.12 Graphical representation of Example 4

QUESTION 1

Evaluate

$$\begin{aligned} x_1-3x_2 &\leq 3\\ x_1-2x_2 &\leq 4\\ x_1 &\leq 8\\ 2x_1+x_2 &\leq 20 \end{aligned}$$

$$x_1+3x_2 \leq 30$$
$$-x_1+\ x_2 \leq\ 6$$
$$2x_1+\ x_2 = \max$$
$$x_1 \geq 0 \qquad x_2 \geq 0$$

using the Simplex tableau, and confirm the result graphically.

QUESTION 2

Attempt to evaluate

$$x_1-\ x_2 \leq 2$$
$$x_1-2x_2 \leq 1$$
$$2x_1+\ x_2+3 = \max$$

Start at $x_1 = x_2 = 0$. What has gone wrong with the solution? Sketch the graph and decide.

QUESTION 3

Two sludge pumps, mounted side by side on the same frame, transfer mud from a settling tank to two hoppers. A single motor drives both pumps through a reduction gear box having a 2 : 1 reduction ratio and in turn through two independent gearboxes having ratios of 4 : 1 and 3 : 1; the arrangement is shown diagramatically in Figure 6.13. There is a danger that solids trapped in the mud could cause the

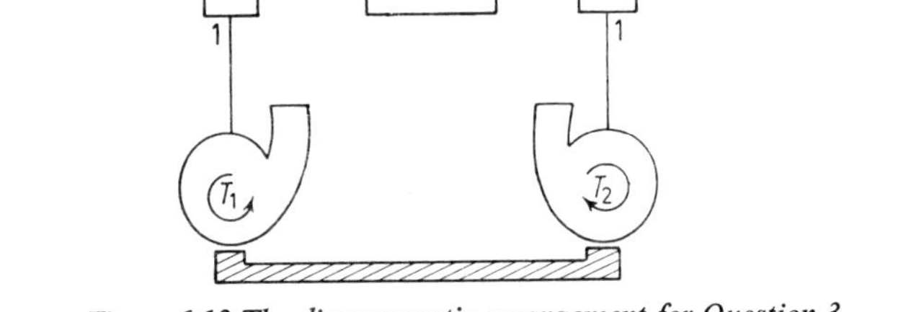

Figure 6.13 The diagrammatic arrangement for Question 3

pumps to lock; to avoid damage to the transmission, the design incorporates a torque limiter on the drive shaft from the motor. The available 2 : 1 reduction gearbox is more than adequate for the task, but considerable cost savings can be made on the other two gearboxes if the input torques can be limited to those shown in the diagram. The strength of the supporting structure restricts the combined torque of the two pumps, T_1 and T_2, to 900 N m. It can be assumed that the pumps are always under load.

Determine, by graphical means, the setting of the torque limiter if these design constraints are not to be exceeded.

Answer

$$\text{Torque setting} = 143 \text{ N m}$$

6.3 DISTRIBUTION APPLIED TO NETWORKS

At times designers are faced with the problem of planning the most economic arrangement for distributing a commodity, e.g. power, fuel, water. When a number of sources exist and they supply several different consumption points the problem of arriving at the optimum arrangement can be complex. The problem is simplified if a linear relationship exists between the quantity being distributed and the cost of distribution, and this will be assumed here. This is a peculiar example of linear programming, which in general can be solved in a much simpler manner. Cost is taken in the widest sense as some form of penalty which has to be minimised, e.g. weight, loss of heat, drop in pressure.

The distribution problem is best illustrated by the following example.

An airline having fifteen aircraft has discovered a design fault on a generator. It is necessary therefore to replace one generator on each aircraft. The aircraft have been grounded in different places but the airline has three service depots distributed about the routes. It is vital to have the aircraft flying as soon as possible.

Table 6.1 TIME IN HOURS

	A_1	A_2	A_3	A_4
D_1	13	11	15	20
D_2	17	14	12	13
D_3	18	18	15	12

The depots are D_1, D_2, and D_3 and each has 2, 6, and 7 replacements respectively. Airfields A_1, A_2, A_3, and A_4 have 3, 3, 4, and 5 aircraft grounded.

The problem is one of distributing the replacements in the least *total* time. *Table* 6.1, gives the time taken to modify, transport, and fit the generators between the various depots and airfields.

Let a typical depot be D_i and an airfield A_j and the number of generators sent from D_i to A_j be x_{ij}, then the units available at the depots have to be distributed to the airfields:

$$x_{11}+x_{12}+x_{13}+x_{14} = 2$$
$$x_{21}+x_{22}+x_{23}+x_{24} = 6$$
$$x_{31}+x_{32}+x_{33}+x_{34} = 7$$

and from the airfield requirements:

$$x_{11}+x_{21}+x_{31} = 3$$
$$x_{12}+x_{22}+x_{32} = 3$$
$$x_{13}+x_{23}+x_{33} = 4$$
$$x_{14}+x_{24}+x_{34} = 5$$

For the aircraft to be out of service the least amount of time, the total generator hours in modification, distribution and fitting must be minimised or, mathematically,

$$13x_{11}+11x_{12}+15x_{13}+20x_{14}+\ldots+12x_{34} = \text{min}$$

i.e. this is the objective function. The problem could be solved using the Simplex method but we will handle it by the transportation or distribution method as follows.

Construct the following table:

	Numbers available	A_1	A_2	A_3	A_4	
		3	3	4	5	*Numbers required*
D_1	2	x_{11}	x_{12}	x_{13}	x_{14}	
D_2	6	x_{21}	x_{22}	x_{23}	x_{24}	
D_3	7	x_{31}	x_{32}	x_{33}	x_{34}	

Considering the cost (time) *Table* 6.1, a reasonable distribution would be to:

take 2 from D_1 to A_2, i.e. $x_{12} = 2$

take 4 from D_2 to A_3, i.e. $x_{23} = 4$

take 5 from D_3 to A_4, i.e. $x_{34} = 5$

filling the table as follows:

		A_1	A_2	A_3	A_4
		3	3	4	5
D_1	2	.	2	.	.
D_2	6	.	.	4	.
D_3	7	.	.	.	5

There are still two generators available at both D_2 and D_3 and deliveries have to be completed at A_1 and A_2. The lowest cost element in this group is D_2A_2, A_2 requiring one unit to complete the order. If this unit is allocated from D_2 then A_1 must receive one unit from D_2 and two units from D_3 giving the following distribution:

		A_1	A_2	A_3	A_4
		3	3	4	5
D_1	2	.	2	.	.
D_2	6	1	1	4	.
D_3	7	2	.	.	5

or $x_{12} = 2$, $x_{21} = x_{22} = 1$, $x_{23} = 4$, $x_{31} = 2$, $x_{34} = 5$, all other $x_{ij} = 0$.

This might appear to be the distribution giving the shortest overall out-of-service time. The total generator hours out-of-service time (or cost) will be

$$2\times 11+1\times 17+1\times 14+4\times 12+2\times 18+5\times 12 = 197 \text{ (generator hours)}$$

Note that the column totals equal the requirements and the row totals equal the availabilities.

If the distribution is rearranged so that $x_{11} = 1$, then 1 must be subtracted from x_{12}, 1 added to x_{22}, and 1 subtracted from x_{21} giving:

		A_1	A_2	A_3	A_4
		3	3	4	5
D_1	2	1	1	.	.
D_2	6	.	2	4	.
D_3	7	2	.	.	5

Observe the − + − sequence.

The total cost of this arrangement is:

$$1\times 13+1\times 11+2\times 14+4\times 12+2\times 18+5\times 12 = 196 \text{ (generator hours)}$$

an improvement of 1, or, alternatively, taking the unit cost per route as C_{ij}, the cost change due to the one-unit rearrangement is:

$$C_{11}-C_{12}+C_{22}-C_{21} = 13-11+14-17 = -1$$

It is still not known whether this is the best overall arrangement.

6.3.1 The Stepping-Stone Method

Every element in the distribution table which contains a value, i.e. every $x_{ij} \neq 0$, is known as a *stepping stone*, and the elements containing no values are known as *waters* ($x_{ij} = 0$).

Remembering that the column and row totals must remain the same, it is possible to see the effective cost change resulting from placing one unit in any of the waters; such a change causes a unique rearrangement in the table.

The above operation can be formalised in a simple mechanical way, being general to any distribution tableau:

(i) Consider each vacant water square in turn. Trace out a path starting at that square, and step vertically or horizontally to stones; a path can be traced out which returns to the water square in question.

(ii) Consider the stones which mark changes between horizontal and vertical movements. Mark the first of these stones with a negative sign and subsequent stones with alternating signs.

(iii) Cost the move of one unit around this path, i.e. assign the sign inserted in the stone to its cost and find the algebraic sum for the path. If this is negative an improvement can be made by a transfer around this path.

(iv) To improve the path, look for the negative stone along the path having the smallest magnitude in x, this is then set equal to zero and the value it contained is subtracted from all the other negatively marked stones on the path, and added to all the positively marked path stones.

(v) The method is systematically repeated for every vacant water in the tableau until no negative algebraic path sums exist.

Instead of making the initial allocation by estimating an economic distribution it is more convenient, particularly with larger problems, to satisfy the distribution by starting the allocation at the top left-hand corner of the tableau and systematically working diagonally across it. It may not be the most efficient way of establishing the initial distribution but because of the rapid convergence of the stepping-stone method it is more convenient. This method is known as the *North-West-Corner Allocation.*

The problem will be reworked employing the north-west-corner allocation and the systematic evaluation. There are many forms of tableau for the distribution method, and the following will be adopted for clarity:

From \ To	A_1	A_2	A_3	A_4	Available
D_1	−[13] 2	+[11]	[15]	[20]	2
D_2	+[17] 1	−[14] 3	[12] 2	[13]	6
D_3	[18]	[18]	[15] 2	[12] 5	7
Requirements	3	3	4	5	

The numbers in the boxes are the times (costs) for the routes. Note that the element $D_1\ A_1$ is the first element to be filled and that the total requirement for A_1 is met from D_2, the remainder at D_2 is allocated first to A_2 and then to A_3; this is the pattern of the north-west-corner allocation.

Consider the water $D_1\ A_2$; the arrows on the tableau show the path associated with that water.

The cost change/unit is:

$$-13+17-14+11 = 1$$

The total is positive, therefore there is nothing to be gained in filling that particular water.

Consider the path for water $D_1\ A_3$:

From \ To	A_1	A_2	A_3	A_4	Available
D_1	− [13] 2	[11]	+ [15]	[20]	2
D_2	+ [17] 1	[14] 3	− [12] 2	[13]	6
D_3	[18]	[18]	[15] 2	[12] 5	7
Requirements	3	3	4	5	

Cost change/unit:

$$-13+17-12+15 = 7$$

Again no improvement results from filling $D_1\ A_3$.

A more complex path exists for $D_1\ A_4$:

From \ To	A_1	A_2	A_3	A_4	Available
D_1	− [13] 2	[11]	[15]	+ [20]	2
D_2	+ [17] 1	[14] 3	− [12] 2	[13]	6
D_3	[18]	[18]	+ [15] 2	− [12] 5	7
Requirements	3	3	4	5	

Cost change/unit:

$$-13+17-12+15-12+20 = 15$$

Systematically considering other waters:

$$D_2\,A_4: \qquad -12+15-12+13 = \ \ 4$$

$$D_3\,A_1: \qquad -15+12-17+18 = -2$$

An improvement can be made by filling D_3A_1, continuing the step-by-step method, i.e. (iv). The negative stone having the smallest magnitude for x is the stone $D_2\,A_1$, in which $x_{21} = 1$. Setting $x_{21} = 1-1 = 0$,

$$x_{23} = 2+1 = 3, \quad x_{33} = 2-1 = 1, \quad \text{and} \quad x_{31} = 0+1 = 1.$$

The improved tableau becomes:

From \ To	A_1	A_2	A_3	A_4	*Available*
D_1	2 [13]	[11]	[15]	[20]	2
D_2	[17]	3 [14]	3 [12]	[13]	6
D_3	1 [18]	[18]	1 [15]	5 [12]	7
Requirements	3	3	4	5	

Note that the other waters and stones remain unchanged and that the column and row totals are satisfied.

Continuing the search for improvement:

$$D_3\,A_2: \qquad -15+12-14+18 = 1$$

again no improvement possible.

The direction around the path is unimportant, as the algebraic total is independent of direction.

Considering the waters once again:

$$D_1\,A_2: \qquad -14+12-15+18-13+11 = -1$$

Set: $x_{33} = 1-1$, $x_{12} = 0+1 = 1$, $x_{22} = 3-1 = 2$, $x_{23} = 3+1 = 4$,

$$x_{31} = 1+1 = 2, \quad \text{and} \quad x_{11} = 2-1 = 1$$

giving:

From \ To	A_1	A_2	A_3	A_4	Available
D_1	[13] 1	[11] 1	[15]	[20]	2
D_2	[17]	[14] 2	[12] 4	[13]	6
D_3	[18] 2	[18]	[15]	[12] 5	7
Requirements	3	3	4	5	

$$D_1 A_3: \quad -12+14-11+15 = 6$$

$$D_1 A_4: \quad -12+18-13+20 = 13$$

$$D_2 A_1: \quad -13+11-14+17 = 1$$

$$D_2 A_4: \quad -12+18-13+11-14+13 = 3$$

$$D_3 A_2: \quad -18+13-11+18 = 2$$

$$D_3 A_3: \quad -18+13-11+14-12+15 = 1$$

i.e. no further improvement is possible.

Note: There is always an even number of terms in the 'cost change/unit' expression.
The total cost is now:

$$1\times 13+1\times 11+2\times 14+4\times 12+2\times 18+5\times 12 = 196 \text{ (generator hours)}$$

There is a number of alternative ways of solving the distribution problem with basically the same manipulations; one is the modified distribution method which enables an operator handworking the tableau to see more clearly the way to an improved solution. Complex distribution problems are normally performed on a digital computer, in which case the user is not concerned with the details of the manipulations but needs to know how to order the input and how to interpret the output.

Summarising:

(i) The north-west-corner allocation is only used for convenience.

6.3.2 The Mathematical Representation of the Distribution Problem

Using the notation already adopted, the problem becomes: minimise the function

$$\sum_i \sum_j C_{ij} x_{ij}$$

subject to $\sum_j x_{ij} = a_i$ and $\sum_i x_{ij} = b_j$

where a_i is the amount available at source i and b_j is the amount required at sink j. Alternatively, for the problem to be feasible the total available at the sources must equal the total requirements at the sinks.

This however does not preclude situations where there is an inequality between the sources and sinks but requires skill by the user in fooling the system to believe that equality exists.

Consider the total output from the sources exceeding the absorption of the sinks. This means an accumulation at the sources, but it is still necessary to optimise the distribution which does take place. This may be handled by introducing a dummy sink, so that the total supply and demand is equal. If unit costs of distribution between the dummy sink and all the sources are 0, then preference will be given to supplying this sink, i.e. retaining the surplus at the source rather than distributing it unnecessarily.

In the reverse situation a dummy source can be introduced where the costs of distribution between the dummy source and the sinks are representative of the penalties in not supplying the sinks. Further information on these inequalities can be found in any good book on operational research.

The distribution problem is a particular case of linear programming; it may be recognised by the fact that all the coefficients of x in the constraint equations must be ± 1, and each x_{ij} must have at the most a $+1$ coefficient and a -1 coefficient between the equations. It would appear that this is not the case, but since the final four equations in the example represent the sinks, these equations can be multiplied throughout by -1.

A limitation arises here. It has really been stated that a route must exist between every source and sink, otherwise some of the coefficients of x will be zero. If a route x_{ij} does not exist there is one of two possibilities; either set the coefficient of x_{ij} to zero, when it must be solved by linear programming which increases the complexity, or the route can be included at an infinite cost. Neither the computer nor the human operator likes infinite numbers, so a route cost is selected which is several orders of magnitude higher than any of the others.

Summarising:

(i) The transportation problem is a particular case of linear programming which can be solved more conveniently by the stepping-stone method.

(ii) There must be equality between sources and sinks but, where there is not, a modified method can be used.

6.3.3 Examples of Network Optimisation

EXAMPLE 5

Four compressors are situated in different parts of a factory and supply five workshops with compressed air. The pipe runs, in metres, between the compressors and the central distribution point in each workshop are given in *Table* 6.2.

Table 6.2

	W_1	W_2	W_3	W_4	W_5
C_1	4	120	160	260	200
C_2	20	90	250	230	220
C_3	200	150	30	90	280
C_4	180	250	240	200	100

The compressors C_1, C_2, C_3, and C_4 produce 4, 3, 6 and 2 m^3/min and the workshops $W_1 \ldots W_5$ require 2, 1·5, 9, 1·2, and 1·3 m^3/min respectively.

The cost of pipe is approximately directly proportional to its flow area. The problem is to establish the distribution network resulting in the minimum cost.

Solution

Let the compressor units be C_1, C_2, C_3 and C_4, or generally C_i, and the workshop distribution points be $W_1 \ldots W_5$, or generally W_j. Let the quantity flow between C_i and W_j be x_{ij}. The sum of quantities flowing between source and sink must equal the availability at source:

$$x_{11}+x_{12}+x_{13}+x_{14}+x_{15} = 4$$
$$x_{21}+x_{22}+x_{23}+x_{24}+x_{25} = 3$$
$$x_{31}+x_{32}+x_{33}+x_{34}+x_{35} = 6$$
$$x_{41}+x_{42}+x_{43}+x_{44}+x_{45} = 2$$

and, from the demand at the workshops,

$$x_{11}+x_{21}+x_{31}+x_{41} = 2$$
$$x_{12}+x_{22}+x_{32}+x_{42} = 1{\cdot}5$$
$$x_{13}+x_{23}+x_{33}+x_{43} = 9$$
$$x_{14}+x_{24}+x_{34}+x_{44} = 1{\cdot}2$$
$$x_{15}+x_{25}+x_{35}+x_{45} = 1{\cdot}3$$

and the objective function becomes

$$4x_{11}+120x_{12}+160x_{13}+260x_{14}\ldots+240x_{43}+200x_{44}+100x_{45} = \text{min}$$

Initial tableau:

From \ To	W_1	W_2	W_3	W_4	W_5	*Available*
C_1	$\boxed{4}$ 2	$\boxed{120}$ 1·5	$\boxed{160}$ 0·5	$\boxed{260}$	$\boxed{200}$	4
C_2	$\boxed{20}$	$\boxed{90}$	$\boxed{250}$ 3	$\boxed{230}$	$\boxed{220}$	3
C_3	$\boxed{200}$	$\boxed{150}$	$\boxed{30}$ 5·5	$\boxed{90}$ 0·5	$\boxed{280}$	6
C_4	$\boxed{180}$	$\boxed{250}$	$\boxed{240}$	$\boxed{200}$ 0·7	$\boxed{100}$ 1·3	2
Requirements	2	1·5	9	1·2	1·3	

Cost change/unit for C_1W_4: $-\ 90+30-160+260 = 40$

Cost change/unit for C_1W_5: $-100+200-90+30-160+200 = 80$

Cost change/unit for C_2W_1: $-250+160-4+20 = -74$

An improvement can be made; the new tableau becomes:

From \ To	W_1	W_2	W_3	W_4	W_5	Available
C_1	4	120 (1·5)	160 (2·5)	260	200	4
C_2	20 (2)	90	250 (1)	230	220	3
C_3	200	150	30 (5·5)	90 (0·5)	280	6
C_4	180	250	240	200 (0·7)	100 (1·3)	2
Requirements	2	1·5	9	1·2	1·3	

C_2W_2: $-120+160-250+90 = -120$

From \ To	W_1	W_2	W_3	W_4	W_5	Available
C_1	4	120 (0·5)	160 (3·5)	260	200	4
C_2	20 (2)	90 (1)	250	230	220	3
C_3	200	150	30 (5·5)	90 (0·5)	280	6
C_4	180	250	240	200 (0·7)	100 (1·3)	2
Requirements	2	1·5	9	1·2	1·3	

C_2W_3: $-90+120-160+250 = 120$

C_2W_4: $-90+30-160+120-90+230 = 40$

C_2W_5: $-100+200-90+30-160+120-90+220 = 130$

C_3W_1: $-20+90-120+160-30+200 = 280$

C_3W_2: $-120+160-30+150 = 160$

C_3W_5: $-100+200-90+280 = 290$

C_4W_1: $-20+250-30+90-200+180 = 270$

C_4W_2: $-120+160-30+90-200+250 = 150$

C_4W_3: $-30+90-200+240 = 100$

C_1W_1: $-120+90-20+4 = -46$

From \ To	W_1	W_2	W_3	W_4	W_5	*Available*
C_1	4 0·5	120	160 3·5	260	200	4
C_2	20 1·5	90 1·5	250	230	220	3
C_3	200	150	30 5·5	90 0·5	280	6
C_4	180	250	240	200 0·7	100 1·3	2
Requirements	2	1·5	9	1·2	1·3	

C_1W_2: $-90+20-4+120 = 56$

C_1W_4: $-90+30-160+260 = 40$

C_1W_5: $-100+200-90+30-160+200 = 80$

C_2W_3: $-20+4-160+250 = 74$

C_2W_4: $-90+30-160+4-20+230 = -6$

From \ To	W_1	W_2	W_3	W_4	W_5	*Available*
C_1	4 1	120	160 3	260	200	4
C_2	20 1	90 1·5	250	230 0·5	220	3
C_3	200	150	30 6	90	280	6
C_4	180	250	240	200 0·7	100 1·3	2
Requirements	2	1·5	9	1·2	1·3	

C_2W_5: $-100+200-230+220 = 90$

C_3W_1: $-4+160-30+200 = 326$

C_3W_2: $-90+20-4+160-30+150 = 206$

C_3W_4: $-30+160-4+20-230+90 = 6$

C_3W_5: $-100+200-230+20-4+160-30+280 = 276$

C_4W_1: $-20+230-200+180 = 190$

C_4W_2: $-90+230-200+250 = 190$

C_4W_3: $-160+4-20+230-200+240 = 94$

C_1W_2: $-90+20-4+120 = 46$

C_1W_4: $-230+20-4+260 = 46$

C_1W_5: $-100+200-230+20-4+200 = 86$

C_2W_3: $-20+4-160+250 = 74$

C_2W_5 This completes the search without producing a further change in the tableau.

The optimum distribution is shown diagramatically in Figure 6.14.

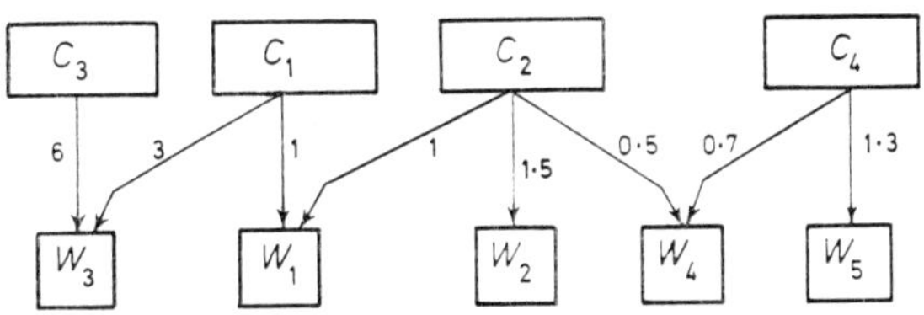

Figure 6.14 The optimum distribution of air between compressors and workshops

Although handworking such a problem is tedious, it should be borne in mind that, on a project of this kind, wastage due to not finding the optimum arrangement could be of the order of several hundred pounds compared with one or two hours of handworking.

Note that the final tableau contains stones in low cost squares and waters in those of high cost. In calculating the cost change per unit transfer it is only necessary to ascertain whether or not the result is positive or negative; this is frequently done by a simple inspection without calculating the value.

EXAMPLE 6

Three boilers supply steam to five turbines. A piping arrangement has to be designed so that minimum thermal energy is contained within the pipes when the turbines are operating at full load. *Table* 6.3 shows the steam output from each boiler and *Table* 6.4 shows the steam required by each turbine. *Table* 6.5 lists the pipe-run lengths between each boiler and turbine.

Write the distribution equations describing the design and set up the initial distribution tableau for the problem.

Table 6.3

Boiler	*Steam available* (kg/s)
1	5
2	4
3	2

Table 6.4

Turbine	*Steam required* (kg/s)
1	1
2	3
3	2
4	2
5	1

Table 6.5

Boiler \ *Turbine*	1	2	3	4	5
1	20	20	60	80	120
2	70	30	20	30	75
3	120	80	60	25	40

Solution

In order to deal with the inequality, a dummy turbine must be introduced. The length of pipe between this turbine and all the boilers must take the value zero, i.e. if the pipe length between a boiler and a turbine is very large, preference will be given to the dummy turbine, which means that the steam will remain at the boiler. Giving the dummy turbine number 6 with suffix *i* denoting the boiler and *j* the turbine and x_{ij} the steam flowing then, for boilers,

$$x_{11}+x_{12}+x_{13}+x_{14}+x_{15}+x_{16} = 5$$

$$x_{21}+x_{22}+x_{23}+x_{24}+x_{25}+x_{26} = 4$$

$$x_{31}+x_{32}+x_{33}+x_{34}+x_{35}+x_{36} = 2$$

For turbines:

$$x_{11}+x_{21}+x_{31} = 1$$

$$x_{12}+x_{22}+x_{32} = 3$$

$$x_{13}+x_{23}+x_{33} = 2$$
$$x_{14}+x_{24}+x_{34} = 2$$
$$x_{15}+x_{25}+x_{35} = 1$$
$$x_{16}+x_{26}+x_{36} = 2 \quad \text{difference between supply and demand}$$

The modified cost table is:

Boiler \ Turbine	1	2	3	4	5	6
1	20	20	60	80	120	0
2	70	30	20	30	75	0
3	120	80	60	25	40	0

and the initial tableau is:

From \ To	T_1	T_2	T_3	T_4	T_5	T_6	Available
B_1	[20] 1	[20] 3	[60] 1	[80]	[120]	[0]	5
B_2	[70]	[30]	[20] 1	[30] 2	[75] 1	[0]	4
B_3	[120]	[80]	[60]	[25]	[40]	[0] 2	2
Requirements	1	3	2	2	1	2	

This tableau as it stands is insoluble since there is a water at x_{35}. By redistributing this part of the allocation to fill the water thus: $x_{35} = x_{25} = x_{26} = \frac{1}{2}$, and $x_{36} = 1\frac{1}{2}$ a solution can be found giving $x_{11} = 1$, $x_{12} = 3$, $x_{13} = x_{16} = \frac{1}{2}$, $x_{23} = 1\frac{1}{2}$, $x_{24} = 1$, $x_{26} = 1\frac{1}{2}$ $x_{34} = x_{35} = 1$.

EXAMPLE 7

If in Example 5 the quality of air available from compressor C_2 is not suitable for use in W_2 (possibly because it is contaminated with oil and the workshop contains pneumatic precision measuring instruments), what would the optimum distribution be?

Solution

To cater for this forbidden route an artificial and extremely high cost is introduced between C_2 and W_2. This is done by taking a very long length of pipe run as illustrated by the modified cost *Table* 6.6.

Note that the objective function of Example 5 will be modified by the new path length.

Table 6.6

	W_1	W_2	W_3	W_4	W_5
C_1	4	120	160	260	200
C_2	20	*1000*	250	230	220
C_3	200	150	30	90	280
C_4	180	250	240	200	100

The initial tableau now becomes:

From \ To	W_1	W_2	W_3	W_4	W_5	*Available*
C_1	[4] 2	[120] 1·5	[160] 0·5	[260]	[200]	4
C_2	[20]	[1000]	[250] 3	[230]	[220]	3
C_3	[200]	[150]	[30] 5·5	[90] 0·5	[280]	6
C_4	[180]	[250]	[240]	[200] 0·7	[100] 1·3	2
Requirements	2	1·5	9	1·2	1·3	

The tableau is now examined as described in Example 5. The final distribution should exclude the route between C_2 and W_2; if this is not the case the artificial length of this route must be raised from 1000 m to a higher figure and the examination repeated.

An alternative approach would be to set the coefficient of x_{22}, as it appears in the solution to Example 5, to zero. The equations are then no longer soluble by the distribution method and must be solved by the simplex method, which may lead to complexity.

6.4 OPTIMISING NONLINEAR RELATIONSHIPS

So far the discussion has been limited to the optimisation of linear relationships, but many engineering systems behave in a nonlinear manner. Unfortunately, there is no standard method of solving nonlinear optimisation problems and satisfactory solutions are as much a matter of experience as analysis. Optimisation of nonlinear systems is grouped under the general headings of: *hill climbing*, this title will become apparent later. Hill-climbing will be described but, because of its complexity, examples will not be attempted; the reader will find a more detailed explanation in references 24 and 25.

Consider a typical car engine. Adjustments are provided in order that the user may obtain the maximum performance from the engine, e.g. mixture adjustment, advance and retard of the spark, contact breaker clearance, sparking plug gap, valve clearances. Suppose that the optimum settings for everything except the spark timing and the mixture have been established and that the development engineer is trying to find these two settings to produce the maximum power under certain operating conditions. It might seem that the correct way of going about the task would be to adjust the mixture; having found the optimum mixture to then adjust the timing to find its optimum, and the job would be completed. More thought would indicate that the mixture would require a further adjustment to compensate for the timing change. In other words the optimum can only be found by adjusting each in turn repeatedly until the variation in the power increase is so small that it becomes immeasurable and here the maximum power is reached.

Figure 6.15 represents this relationship in a graphical form. The two axes are in units of timing and mixture ratio and the power axis is normal to the paper surface. The contour lines are lines of constant power and are similar to those on a map showing the contour of a hill.

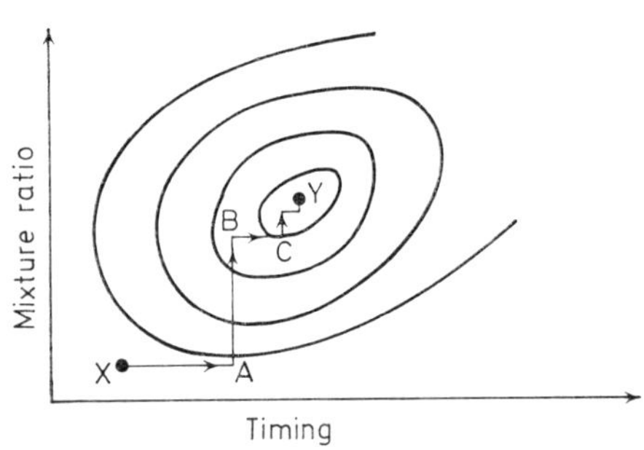

Figure 6.15 The graphical representation of the mixture ratio and timing adjustment of an engine

The 'peak' or optimum value of the power is at Y. If the search for the optimum started at X the first adjustment to timing would produce a change to A. The first adjustment of the mixture would move the search to B and the second adjustment of timing to C and so on until the maximum had been reached. Note that the changes decrease in magnitude as the optimum is approached. This form of hill climbing is known as a *direct-search* method. Consider the contours of the hill in Figure 6.16(a). If the search is started at X the climb will soon begin to follow the ridge, and if the ridge should be very narrow the increments in x_1 and x_2 will become very small and may become immeasurable before the maximum is reached. If the start is at Y the

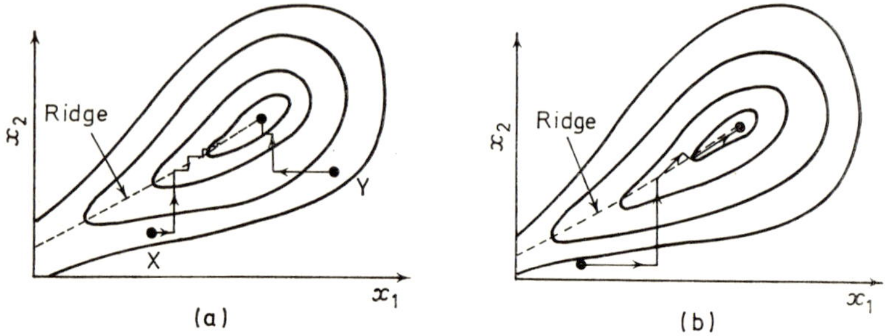

Figure 6.16 The effect of a ridge in the region of search

problem will not be encountered. Figure 6.16(b) shows a method of overcoming this difficulty, attributed to Powell[26]. The search is started as before but for the third step of the search the search direction is aligned with the starting point and the fourth step is made at right angles to the third. This means that every odd step of the search now becomes approximately aligned with the ridge which the search is following.

It may seem logical that the quickest way to proceed to the peak would be to stand at a point and examine the first derivative of the surface to find the direction of steepest gradient, by following this direction the peak would be reached more rapidly. Such methods are known as *gradient* methods for obvious reasons. The difficulty is that a known functional relationship must exist for the first derivative to be calculated. The tuning of the car engine illustrates one case where the functional relationship is unknown, which is probably true in the majority of cases. If a sufficiently large number of values for the function exist it may be possible to define the hill by a surface equation but, although the surface may match the function almost exactly, the derivatives arrived at in this way could be grossly in error. As a consequence, direct search methods are more generally applied.

A further difficulty encountered in nonlinear optimisation is knowing whether the maximum value found is in fact the real maximum. Figure 6.17 shows two maxima occurring within the region of search, X being less than Y. If the search is started at B the true maximum will be found, whereas a search started at A would converge on the false value. As a consequence, no nonlinear optimisation method guarantees that an optimum has been found. Nonlinear optimisation is normally employed where the terrain is fully understood. In this way a computer can be programmed to seek out the true optimum of the region; hence the extensive use of this technique

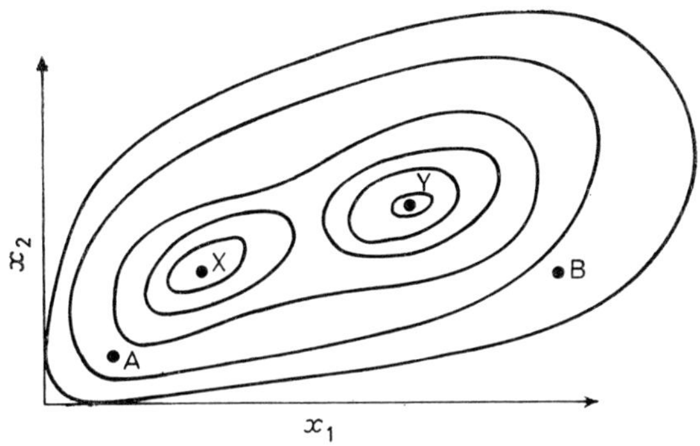

Figure 6.17 The problem of two maxima being within the region of search

for the optimum control of chemical plant where the process variables can be fully explored before the control is set up.

Arguments so far have been limited to two independent variables, but the same techniques can be extended to several variables, although typical limits might lie between 5 and 20 because of complexity.

Although nonlinear optimisation is an essential consideration in many design problems, it is still the field of the specialist and offers an obvious incentive to use linear approximations wherever possible.

Summarising:

(i) There is no guarantee with the hill-climbing method that the true maximum or minimum will be found.

(ii) Its use is limited to only a few variables.

Appendix

Transportation

Sea Freight

The maximum permitted sizes and weights vary since they are dependent upon the port facilities and on the type of fleet operated.

The standard equivalent volume is:

40 ft^3 *represents* 1 ton

The volume is calculated from the extreme points of the three dimensions, excluding lifting aids such as hooks, eyes and lugs, hence a design needs to eliminate unnecessary projections during transit. As a result, bulky articles are charged on volume, 40 ft^3 being costed as 1 ton.

SYSTEMS OF RATING

Company	*Normal rate*	*Maximum standard size*	*Exceeding standard size*	*Maximum standard dimension*	*Exceeding standard dimension*
A	Depends on commodity	5 tons	Surcharge	Max. length 30 ft 2 in	Surcharge
B	Normally independent of commodity	2 tons	Rated on contents	Large loads to be negotiated	Rated as a special project
C	Depends on commodity	3 tons	Warnings of shipment required	Max. height 6 ft 6 in	Warning of shipment required

Where the rate depends on contents it is usually based on such things as commodity value, difficulty of handling, e.g. having to keep a container a certain way up, etc.

Container Freight

Container sizes:

(i) *small*	8 ft×8 ft×8 ft contents up to 4 tons (mainly short sea haul)
(ii) *medium*	20 ft×8 ft×8 ft contents up to 18 tons
(iii) *large*	40 ft×8 ft×8 ft contents up to 31 tons

Note: Some countries have a maximum load limit for road-borne container traffic. In the UK this is limited to about 23 tons after making an allowance for the trailer and motive unit.

Road Haulage

In the UK the usual haulage density is 80 ft^3 equivalent to one ton, or 2 ft length of trailer space is equivalent to 1 ton. *Sizes:* The only limits on height are overhead obstructions encountered en route, e.g. bridges, and the handling, i.e. the centre of gravity of the load and transporter must be low enough to be stable.

In the UK the width above which the police must be notified for routeing is 9 ft 6 in.

In the UK the load is permitted to overhang the trailer front and rear by 6 ft, giving an additional twelve feet in length.

Rail Freight

British Rail Express Parcel Service:

Maximum weight 112 lb
Maximum standard length 7 ft
Maximum length 18 ft 6 in (surcharge made for items between 7 ft and 18 ft 6 in)
Maximum standard density 300 ft^3/ton (extending to 2000 ft^3/ton with approximately a 200% surcharge)

Air Freight

Temperature variations encountered:	−40 °C to 54 °C
Pressure variations encountered:	15 lbf/in^2 to 11 lbf/in^2, or sometimes 6 lbf/in^2

Trapped gases in containers should be capable of withstanding an additional pressure difference of 1 atmosphere. Air must be removed from all sealed wrappings. Standard density: up to 7000 cm^3/kg.

Restricted articles: magnetised material, explosives, radioactive materials, poisons, flammable materials, compressed gas, corrosive liquids, etc.

Publications: A.B.C. Air Cargo Guide, published by Thomas Skinner and Co. Ltd, 30 Finsbury Square, London EC2 (gives details of airline routes, commodity descriptions, etc.)

I.A.T.A. Bulk Unitisation, published by I.A.T.A., 1155 Mansfield Street, Montreal.

I.A.T.A. Containers: density up to $1\frac{1}{2}$ lb/ft^3 (container weight not charged).

Standard sizes: (see Figure A.1—dimensions in cm)

Size number	*Length* (cm)	*Width* (cm)	*Height* (cm)	*External volume* (m^3)
1	214	259	193/168/115*	10·154
2	214	303	193/115*	11·513
3	214	259	115	6·374
4	214	303	115	7·457
5	214	148	193/115*	5·645
6	214	107	193/168*	4·379
7	214	148	115	3·642
8	214	102	115	2·510
9	107	148	115	1·821
10	107	127	148	2·011
11	214	132	178	5·028
12	107	132	178	2·514
13	107	132	115	1·624
14	107	132	89	1·257
15	107	102	115	1·255
16	107	74	115	0·911
17	107	74	65	0·515

* Contoured

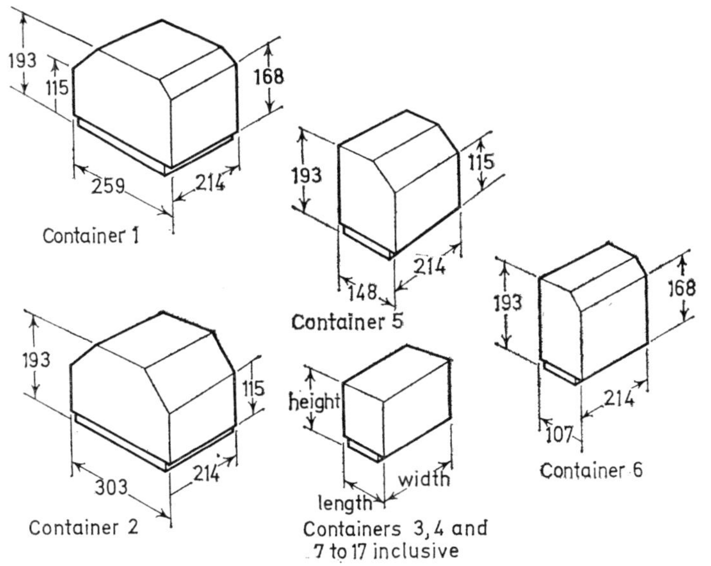

Figure A.1

Fragile Items

The responsibility for fragile items usually lies with the owner of the goods, hence the need for secure packing and good design to facilitate packing.

Conclusions

(i) Products should, in general, be designed for a high packing density.

(ii) Bulk becomes more of a problem when goods are being sent overseas.

(iii) For overseas shipments it is desirable to limit individual packages to less than 2 tons, or to design to make the maximum use of container space and load.

(iv) Large loads which have to be shipped through several countries may be better handled as container traffic.

References

1. PYE, D., *The Nature of Design*, Studio Vista, London (1964)
2. MATOUSEK, R., *Engineering Design: A Systematic Approach*, Blackie, London (1963)
3. HARRISBERGER, L., *Engineersmanship—A Philosophy of Design*, Brook/Cole Pub. Co., Belmont, Calif. (1966)
4. Engineering Sciences Data Unit, London. Various publications
5. LATHAM, R. L., *A Guide to the Problem Analysis by Logical Approach System*, 5th edn, United Kingdom Atomic Energy Authority, AWRE, Aldermaston, Berks (Jan. 1968)
6. TERRY, G. J., 'A Chart System to Help Designers', *Chartered Mech. Engr.*, **15,** No. 2, 56–59 (Feb. 1968)
7. ZWICKY, F., 'The Morphological Method of Analysis and Construction', *Courant*, Anniversary Vol. (1948)
8. NORRIS, K. W., 'The Morphological Approach to Engineering Design', *Proc. Conf. on Design Methods*, Pergamon Oxford (1963)
9. DAVIS, L. A., 'Speeding up Design and Value Engineering, Readers' Views', *Chartered Mech. Engr*, **15,** No. 6, 271 (June 1968)
10. PITTS, G., 'A Design Method for tilting-pad Gas Journal Bearings', *Tribology*, 8–21 (Feb. 1970)
11. ELLIOT GREEN R., 'Computer Graphics', *Computer Aided Design*, suppl. **2,** No. 3, 29 (Spring 1970)
12. DUSINBERRE, G. M., *Heat Transfer Calculations by Finite Differences*, International Textbook Company, Scranton, Pennsylvania (1961)
13. IQBAL, A., 'Computer Aided Design of Storage Heaters', *Proc. IEE Conf. on Computer Aided Design, IEE, London* (Apr. 1972)
14. CALOGERO, V., 'A New Method in Road Design: Polynomial Alignment', *Computer Aided Design*, 19–29 (Winter 1969)
15. AHLBERG, J. H., NILSON, E. N., and WALSH, J. L., *Theory of Splines and Their Applications*, Academic Press, New York (1967)
16. FERGUSON, J. C., 'The Method of Multivariable Curve Interpolation', *J. Ass. Computing Mach.*, 221–228 (Apr. 1964)
17. COONS, S. A., *Surfaces for Computer Aided Design of Space Forms*, MIT (1967)
18. BÉZIER, P., *Emploi Des Machines A Commande Numerique*, Masson et Cie, Paris (1970)

19. SABIN, M. A., 'The British Aircraft Corporation Numerical Master Geometry System', *Proc. R. Soc., Lond.*, A.**321**, 197–205 (1971)
20. ARMIT, A. P., 'Curve and Surface Design Using Multipatch and Multiobject Design Systems', *Computer Aided Design*, **3**, No. 4, 3 (Summer 1971)
21. MISCHKE, C. R., *An Introduction to Computer-Aided Design*, Prentice-Hall, New Jersey (1968)
22. DANTZIG, G. B., *Linear Programming and Extensions*, Univerity of Princeton Press (1963)
23. JOHNSON, R. C., *Optimum Design of Mechanical Elements*, Wiley, New York (1961)
24. ROSENBROCK, H. H., and STOREY, C., *Computational Techniques for Chemical Engineers*, Pergamon, Oxford (1966)
25. DIXON, L. C. W., *Nonlinear Optimisation*, English Universities Press (Feb. 1972)
26. POWELL, M. J. D., 'An Iterative Method for Finding Stationary Values of a Function of Several Variables', *Computer J.*, **5**, 147–151 (1963)

Bibliography

DESIGN METHOD

ASIMOW, M., *Introduction to Design*, Prentice-Hall (1962)

DIXON, J. R., *Design Engineering: Inventiveness, Analysis, and Decision Making*, McGraw-Hill (1966)

FRENCH, M. J., *Engineering Design: The Conceptual Stage*, Heinemann (1971)

GLEGG, G. L., *The Design of Design*, Cambridge University Press (1969)

HILL, P. H., *The Science of Engineering Design*, Holt Rinehart and Winston (1970)

JONES, J. C., *Design Methods: Seeds of Human Futures*, Wiley (Dec. 1970)

WOODSON, T. T., *Introduction to Engineering Design*, McGraw-Hill (1966)

DESIGN ECONOMICS

BUCK, C. H., and BUTLER, D. M., *Economic Product Design*, Collins (Feb. 1971)

CRUM, L. W., *Value Engineering: The Organised Search for Value*, Longman (1971)

OUGHTON, F., *Value Analysis and Value Engineering*, Pitman (1969)

RAVEN, A. D., *Profit Improvement by Value Analysis, Value Engineering and Purchase Price Analysis*, Cassell (1971)

COMPUTERS IN DESIGN

DODD, K. N., *Computer Programming and Languages*, Butterworth, London (1969)

EARNSHAW, J. D. and BLACKFORD, W. A. R., *Computer Programming in Algol*, Pitman (1970)

FURMAN, T. T., Ed., *The Use of Computers in Engineering Design*, English Universities Press (1970)

WOLFENDALE, E., Ed., *Computer-Aided Design Techniques*, Butterworth, London (1970)

NUMERICAL SPLINES & SURFACES

KUO, C., *Computer Methods for Ship Surface Design*, Longman (1971)

PRINCE, M. D., *Interactive Graphics for Computer-Aided Design*, Addison-Wesley (1971)

OPTIMISATION

BEVERIDGE, G. S. G., and SCHECHTER, R. S., *Optimization: Theory and Practice*, McGraw-Hill (1970)

JOHNSON, R. C., *Mechanical Design Synthesis: With Optimization Applications*, Van Nostrand Reinhold (Aug. 1971)

REINFELD, N. V. and VOGEL, W. R., *Mathematical Programming*, Prentice-Hall (1958)

VAJDA, S., *Readings in Mathematical Programming*, Pitman (1962)

ZENER, C., *Engineering Design By Geometric Programming*, Wiley (Nov. 1971)

GENERAL

BODDY, D. E., *Engineering Design Computation Manual*, Holt, Rinehart and Winston (1969)

DANCY, J., *Applied Design '71: Design Engineering Handbook of Applied Design*, Morgan-Grampian (1970)

DUGGAN, T. V., *Applied Engineering Design and Analysis*, Iliffe (1970)

DUGGAN, T. V., *Power Transmission and Vibration Considerations in Design*, Butterworth, London (1971)

EDER, W. E., and GOSLING, W., *Mechanical System Design*, Pergamon (1965)

LARSEN, E., *A History of Invention*, Dent (1969)

MAYALL, W. H., *Industrial Design for Engineers*, Butterworth, London (1967)

PHELAN, R. M., *Fundamentals of Mechanical Design*, McGraw-Hill (1970)

REDFORD, G. D., *Mechanical Engineering Design*, Macmillan (1966)

ROYLANCE, T. F., Ed., *Engineering Design (Proc. Conf. at University of Nottingham)*, Pergamon, Oxford (1966)

SHIGLEY, J. E., *Mechanical Engineering Design*, McGraw-Hill (1963)

TRIBUS, M., *Rational Descriptions, Decisions and Designs*, Pergamon (1969)

Index